T0144077

Artificial Intelligence in a Throughput Model

Some Major Algorithms

Waymond Rodgers
Chair Professor, University of Hull, UK
University of Texas El Paso, Texas, USA

CRC Press
Taylor & Francis Group
Boca Raton London New York

CRC Press is an imprint of the
Taylor & Francis Group, an **informa** business

A SCIENCE PUBLISHERS BOOK

CRC Press
Taylor & Francis Group
6000 Broken Sound Parkway NW, Suite 300
Boca Raton, FL 33487-2742

© 2020 by Taylor & Francis Group, LLC
CRC Press is an imprint of Taylor & Francis Group, an Informa business

No claim to original U.S. Government works

Version Date: 20200120

International Standard Book Number-13: 978-0-367-21781-5 (Hardback)

Visit the Taylor & Francis Web site at
http://www.taylorandfrancis.com

and the CRC Press Web site at
http://www.crcpress.com

Preface

"[Artificial Intelligence] is going to change the world more than anything in the history of mankind. More than electricity."
— AI oracle and venture capitalist Dr. Kai-Fu Lee, 2018

In 1964, Isaac Asimov envisioned the 2014 World's Fair for *The New York Times*. He was correct about the smartphone, self-driving cars and the Keurig machine; however, he was not able to predict advanced battery technology and space colonization. Foresight is challenging, and is only becoming more and more so.

Artificial Intelligence (AI) is the next major disruptor in the way we live, learn, work and adopt to different situations. Disruptors represent a person, place or thing that prevents something, especially a system, process or event, from continuing as usual or as expected. Our houses, our cars, our toasters, all of seem to be teeming, even overflowing with intelligence, like some huge fungus gone amuck. Artificial Intelligence is here to stay, and society needs it, right now! Artificial Intelligence is changing the world each day. At one time, a domain of science fiction, today many of our systems are powered by Artificial Intelligence.

The field of Artificial Intelligence has awed researchers and users alike. Artificial Intelligence is the intelligence of machines and a subset of computer science and cognitive science. Fundamental challenges of Artificial Intelligence embrace such features as reasoning, knowledge, planning, learning, communication, perception and the capability to move and manipulate objects.

This book addresses the promise of Artificial Intelligence that is enhancing our lives. Artificial Intelligence-based systems are now outshining medical specialists in diagnosing certain diseases. The implementation of Artificial Intelligence in the financial system is magnifying access to credit to borrowers whose loan applications were once rejected. In addition, automated hiring systems have the potential to evaluate candidates on the basis of their authentic qualifications as opposed to characteristics such as age or appearance that oftentimes mislead decision makers from making the correct decision.

One of the many advantages of utilizing Artificial Intelligence and machine learning is that it has the ability to ingest huge amounts of data, often in real-time. Further, it can take that data and begin to scrutinize it based upon organizational necessities, conditions and constraints. In addition, it can bring about those necessities, conditions and constraints based on the data an organization owns.

Moreover, many physical and behavioral biometric technologies such as fingerprint, facial recognition, voice identification, etc., have enhanced the level of security substantially. Governments and corporates have embraced these technologies to increase customer satisfaction. Nevertheless, the current state of biometrics still faces challenges to lessen criminal and terrorist activities as well as other digital-based financial frauds. This is especially the case when individuals and organizations are faced with selecting the correct algorithm to a problem. To overcome this state of affairs, the market undertakes a host of research and development programmes to assimilate biometrics with artificial intelligence in decision-making modeling. The advanced software algorithm platform of Artificial Intelligence processes information offered by biometric technology to detect and prevent dubious activities in a bid to confront cyber and physical crimes in the global and local communities. This development has provided an expanded growth opportunity for the biometrics technology, given that the technology is set to increase the security and internal control operations many folds.

This book provides an overview of the various Artificial Intelligence techniques, biometric technologies, decision-making algorithms and the subsequent market expansion opportunities. Further, it proposes a Throughput Model, which draws from computer science, economics and psychology to model perceptual, informational sources, judgmental processes and decision choice algorithms. This approach provides how huge data and biometrics might be implemented to reduce risks to individuals and organizations, especially when dealing with digital-based mediums.

The book also examines the ethics behind Artificial Intelligence. That is, how machine learning, neural networks, and deep learning technology are positioned today for many individual and organizational uses, including self-driving cars, online recommendations, search engines, handwriting recognition, computer vision, online ad serving, pricing, prediction of equipment failure, credit scoring, fraud detection,

OCR (optical character recognition), spam filtering, etc. Therefore, this book addresses ethical consideration directed at the growing ubiquity of machine learning, neural networks, and deep learning in organizations. This particular issue is essential in order to understand what and how to mitigate human cognitive biases and heuristics into Artificial Intelligence technology.

Acknowledgements

First and foremost, my thanks to the Almighty for His blessings throughout the course of my research work without which I would not have been able to complete this research successfully. I also acknowledge deeply all those who have helped me to put these ideas well above the level of simplicity into something concrete.

Further, I would like to thank my students. Learning is a collaborative activity when it is happening at its best. We work together using each other's strengths to build our own challenges, developing our thinking and problem solving skills. Therefore, the relationship we develop with our students at every age is one that is to be respected, nurtured and admired.

Last but not the least, I shall be forever indebted to all those who have been with me throughout the course of this research but whose names I am not mentioning individually.

Contents

1

Introduction to Artificial Intelligence and Biometrics Applications

Artificial Intelligence is the science and engineering of making intelligent machines, especially intelligent computer programs.
                    —John McCarthy, Father of Artificial intelligence

"Man is only great when he acts from passion."
        —Benjamin Disraeli (former Prime Minister of Britain)

Artificial intelligence (AI) is here for stay. For individuals and organizations, Artificial Intelligence is a disruptor in the way we live, learn, work and adopt to different situations. Disruptors represent a person, place or thing that prevents something, especially a system, process, or event, from enduring as customary or as anticipated in the future. We are right in the midst of the information revolution. Although it is an extraordinary time and place to be in, there are caveats that come along with it. Having a machine tell you how long your commute will be, what music you should listen to, and what content you would likely engage with are all relatively innocuous examples.

Artificial Intelligence is presently used as a tool to assist people. It is adopted as point solutions across a wide array of functions such as personal digital assistant, email filtering, search, fraud prevention, engineering, marketing models, digital distribution, video production, news generation, play and game-play analytics, customer service, financial reporting, marketing optimization, energy cost management, pricing, inventory, enterprise applications, etc. Artificial Intelligence is also integrated into biometrics tools such as iris recognition, voice recognition, facial recognition, content classification, gait, and natural language.

In addition, Artificial Intelligence is becoming widespread in most facets of decision-making and will become more so in the near future. Artificial Intelligence has been an aspiration of computer scientists since the 1950s, and has experienced colossal advancement in recent years. Artificial Intelligence implementation is already an integral part of many of our online activities and will become progressively more entrenched in everything we touch, hear, see and do. On a task-by-task basis, Artificial Intelligence systems gradually produce outputs that far exceed the precision and reliability of those produced by individuals. For example, pharmaceuticals and the food industry utilize Artificial Intelligence sensor tools to ensure the optimum temperature for creating drugs or cooking food. Other sensors make certain that products are stored and shipped at safe temperatures.

In agriculture, the implementation of Artificial Intelligence tools has boosted crop yields. Artificial Intelligence has provided for farmers to make better decisions pertaining to crops to sow and how to best manage them. Harvesting everything from grains to root vegetables to soft fruits can now be performed more efficiently and effectively with robots than people. Furthermore, mobile technology has reformed the manner in which field service teams operate. They can get work orders more rapidly, and once the employees are at a particular location, they can access schematics and documents to assist with the repair work. In addition, the Artificial Intelligence system provides for a smoother progression of work orders from fault call to the ordering of parts and the ensuing billing of satisfied customers.

In financial services, PwC has garnered enormous amounts of data from the US Census Bureau, US financial data, and other public licensed sources to create secure, a large-scale model of 320 million US consumers' financial decisions (https://www.pwc.com/gx/en/services/advisory/consulting/security-cyber-assets.html). The model is intended to assist financial services firms map buyers' personas, simulate "future selves" and anticipate customer behavior. It has empowered these financial services companies in substantiating real-time business decisions within seconds.

Similar to the financial services sector, Artificial Intelligence has been implemented to develop a model of the automobile ecosystem. Here, you have bots that map the decisions made from automotive players, such as vehicle purchasers, manufacturers, and transportation services providers. This has assisted organizations to predict the adoption of electric and driverless vehicles, and the enactment of non-restrictive pricing schemes that work on their target market. It has also assisted them in providing enhanced advertising decision choices.

Artificial Intelligence application development has delivered to marketeers with new and more reliable tools of market forecasting, process automation and decision-making (https://www.tenfold.com/business/artificial-intelligence-business-decisions). Further, Artificial Intelligence can be employed while individuals are scrolling through their social media newsfeed, an algorithm somewhere is determining someone's medical diagnoses, their parole eligibility, investment and purchasing habits, or their career prospects.

The remaining chapter sections deal with (1) Artificial Intelligence categories, (2) six Throughput Model algorithms driving Artificial Intelligence, (3) impact of Artificial Intelligence, (4) expert systems and Artificial Intelligence systems (including machine learning, neural networks, deep learning and natural language systems).

Categories of Artificial Intelligence

According to Arend Hintze, an assistant professor of Integrative Biology and Computer Science and Engineering at Michigan State University, Artificial Intelligence can be encapsulated into four classes, from the kind of Artificial Intelligence methods that are in existence today to systems that are yet to come into being (https://searchenterpriseai.techtarget.com/definition/AI-Artificial-Intelligence).

Artificial Intelligence can be defined as (1) "narrow" Artificial Intelligence, (2) "general" (or strong) Artificial Intelligence, or (3) "super" Artificial Intelligence. Narrow Artificial Intelligence is able to conduct just one particular task. Further, this Artificial Intelligence type can attend to a task in real-time; however, they pull information from a specific data-set. As a result, these systems do not execute outside of the single task that they are designed to perform.

Narrow Artificial Intelligence has only characteristics consistent with cognitive intelligence. These Artificial Intelligence systems engender a cognitive representation of the world and utilize learning based on previous experiences to bring up-to-date future decisions. Most Artificial Intelligence systems implemented by today's organizations fall into this group. Examples include systems used for fraud detection in financial services, image recognition, or self-driving cars.

Narrow Artificial Intelligence is not conscious, sentient, or driven by emotions the way that individuals are configured to make decisions. Narrow Artificial Intelligence operates within a pre-determined, pre-defined range, even if it gives the impression to be much more sophisticated than that due to heuristics and biases (to be discussed in later chapters). Every type of machine intelligence that encircles us in the global society is narrow Artificial Intelligence. Examples of narrow Artificial Intelligence include Google Assistant, Google Translate, Siri and other natural language processing tools. These systems lack the self-awareness, consciousness, and genuine intelligence to match human intelligence. That is, they cannot think for themselves. Nonetheless, narrow Artificial Intelligence by itself is a great accomplishment in human innovation and intelligence.

On the other hand, general Artificial Intelligence's are more sophisticated. They are able to cope with any generalized task asked of it, much like a human being. In other words, general Artificial Intelligence's represent machines that display human intelligence, that is, they are able to perform any intellectual task that a human being can in terms of decision-making. This is the kind of Artificial Intelligence that we see in military operations and in movies whereby humans interact with machines and operating systems that are conscious, sentient, and driven by emotions and self-awareness.

General Artificial Intelligence has elements from both cognitive as well as emotional intelligence. Moreover, these systems depict cognitive elements, understand human emotions and include them in their decision making. Affectiva, an Artificial Intelligence firm founded by MIT, uses advanced vision systems to recognize emotions such as joy, surprise, and anger at the same level (and frequently better) as people (https://www.affectiva.com/). Organizations can implement such systems to recognize emotions during customer involvements or while recruiting new employees.

Classes one and two are denoted as "narrow" Artificial Intelligence, whereas, classes three and four are designated as "general" Artificial Intelligence, as follows:

Class 1: Reactive machines. An example is Deep Blue, the IBM chess program, which defeated Garry Kasparov in the 1990s. Deep Blue can recognize pieces on the chess board and generate predictions. Nonetheless, it has no memory and cannot implement previous experiences to update future ones. It scrutinizes possible moves, its own as well as its opponents'. It then selects the next strategic position. Deep Blue and Google's AlphaGo (a Chinese strategy board game) were designed for narrow purposes and cannot simply be employed to another setting.

Class 2: Limited memory. The automotive industry has fostered several Artificial Intelligence applications, from vehicle design to marketing and sales decision-making support. Moreover, Artificial Intelligence has led to the design of driverless automobiles fortified with multiple sensors that learn and identify patterns. This is utilized through add-on safe-drive features that warn drivers of possible collisions and lane departures.

These Artificial Intelligence systems can utilize previous encounters to inform future decisions. Some of the decision-making functions in driverless self-driving automobiles (oftentimes referred to as an autonomous car/driverless car) are designed in this manner. A self-driving car is a vehicle that utilizes a combination of sensors, cameras, radar and Artificial Intelligence in order to travel between destinations without a human operator. To qualify as fully autonomous, a vehicle must be able to navigate without human involvement to a prearranged destination over streets and lanes that have not been modified for its function. Further, observations inform actions occurring in the near future, such as a car changing lanes. These observations are not stored permanently. Moreover, the US National Highway Traffic Safety Administration (NHTSA) (https://whatis.techtarget.com/definition/National-Highway-Traffic-Safety-Administration-NHTSA) has put down six levels of automation, beginning with zero, where humans do the driving, through technologies up to fully autonomous cars. Driver assistance, also known as advanced driver-assistance systems (ADAS), are technologies implemented to craft motor vehicle travel safer by automating, improving or adapting some or all of the tasks involved in operating a vehicle.

The six levels are:

Level 1: Humans do the driving.

Level 2: ADAS supports the human driver with either steering, braking or accelerating, though not simultaneously. ADAS encompasses rearview cameras and characteristics such as a vibrating seat warning to alert drivers when they drift out of the traveling lane.

Level 3: An ADAS that can steer and either brake or accelerate all together while the driver remains fully mindful behind the wheels and continues to act as the driver.

Level 4: An automated driving system (ADS) can perform all driving responsibilities under certain circumstances, for instance, parking the car. In these situations, the human driver must be ready to re-take control and is still expected to be the main driver of the vehicle.

Level 5: An ADS is able to perform all driving tasks and observe the driving environment in certain situations. In those situations, the ADS is reliable enough for the human driver to not have be attentive.

Level 6: The vehicle's ADS acts as a virtual chauffeur and does all the driving in all situations. The human occupants are passengers and are never expected to drive the vehicle.

Class 3: Theory of mind (general Artificial Intelligence). This psychology term denotes the understanding that others have their own wants, desires, needs, beliefs, values and attitudes that impact the choices they make. This type of Artificial Intelligence does not yet exist.

Class 4: Self-awareness (general Artificial Intelligence). In this grouping, Artificial Intelligence systems have a sense of self, and have consciousness. Machines with self-awareness understand their current state and can utilize the information to deduce what others are feeling. This type of Artificial Intelligence does not yet exist.

Based upon the aforementioned types of Artificial Intelligence competencies (i.e., narrow and general), we can add expert systems (discussed later in the chapter) and the yet to come super Artificial Intelligence systems depicted in Table 1.1. An expert system is a computer system that emulates the decision-making capability of an individual expert. Expert systems are devised to explain and decipher complicated problems by reasoning through bodies of knowledge, exemplified primarily as if–then rules rather than through established procedural code (Housel and Rodgers, 1994).

Table 1.1. Advancement of Artificial Intelligence systems

	Expert Systems	Narrow Artificial Intelligence	General Artificial Intelligence	Super Artificial Intelligence	Human Beings
Cognitive Intelligence	x	✓	✓	✓	✓
Emotional Intelligence	x	x	✓	✓	✓
Social Intelligence	x	x	x	✓	✓
Artistic Creativity	x	x	x	x	✓

Partially adapted from Kaplan and Haenlein, 2019.

Super Artificial Intelligence displays features of all kinds of competencies (i.e., cognitive, emotional, and social intelligence). Such systems, which would be able to be self-conscious and self-aware in their interactions with others, are not currently available yet. While progress has been made in recognizing and mirroring human activities, constructing Artificial Intelligence systems that actually experience the world in a basic manner are a project for the (potentially distant) future.

The Six Macro Throughput Model Algorithms Driving Artificial Intelligence

Artificial Intelligence with its companions machine learning, neural networks and deep learning algorithms (discussed in detail later in this chapter) depicts a promising solution for mitigating the nefarious problem of human bias (see Figure 1.1). Also, it can negatively impact the lives of millions of people in various ways. The idea is that the algorithms in

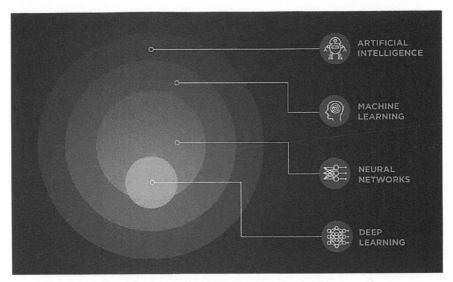

Figure 1.1. Components of Artificial Intelligence

Artificial Intelligence are capable of being fairer and more efficient than humans could ever be in this time period.

For example, deep learning can be employed to tackle issues of scale often prevalent in the execution of organizational schemes. It is essentially a process that can be used for pattern recognition, image analysis and natural language processing (NLP) by modelling high-level abstractions in data which can then be linked with numerous other identifiable contents in a conceptual manner as opposed to exploiting just a rule-based method. The span of application for Artificial Intelligence techniques in such large-scale public endeavors could range from crop insurance schemes, tax fraud detection, detecting subsidy leakage to defense and security strategy.

Profit and nonprofit organizations, governments, and individuals worldwide are integrating Artificial Intelligence empowered decision-making for many reasons. One, it is more reliable and relevant, it becomes easier, it is less costly, and it's time-efficient. Nevertheless, there are still some concerns to be aware of in this era of Artificial Intelligence. Therefore, this book examines a Throughput Model that embraces six dominant cognitive algorithms that are tied to machine learning, neural networks and deep learning. The later chapters will explore the Throughput Model by means of cyber security issues, big data usage and ethical considerations. The Throughput Model developed by Rodgers (1984), posits that four major concepts of perception, information, judgment and choice are implemented in a certain sequence before a decision choice (Figure 1.2). Further, not all the four major concepts are necessary in each of the six

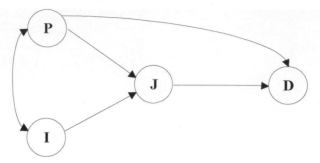

Figure 1.2. Throughput Model: Decision makers' algorithmic processes

dominant cognitive algorithmic pathways (discussed in more detail later in this chapter).

These six cognitive algorithmic pathways are described as follows:

(1) **P → D**
(2) **P → J → D**
(3) **I → J → D**
(4) **I → P → D**
(5) **P → I → J → D**
(6) **I → P → J → D**

Moreover, these six critical algorithmic pathways, separates the decision-making process into its four main parts in different arrangements resulting in an action or no action: perception (**P**), information (**I**), judgment (**J**) and decision choice (**D**) (Rodgers, 1984, 1991a,b, 1997, 2006, 2016). In this model perception and information are interdependent since information can influence how the decision-maker frames a problem (perception) or how he/she select the evidence (information) to be used in later decision-making stages (judgment and choice). The interaction of perception and information can be viewed as a neural network in that people are constantly receiving information and they attempt template match what is in their memory, which begins with perception (i.e., problem framing).

This interdependency is similar to a neural network or LENS model approach (Hammond et al., 1980) which compares individuals' or Artificial Intelligence systems depicting perceptions of information signals with the actual information presented to the individuals as well as updates one's perception with information (i.e., P←→I). Moreover, in the first stage, perception and information influence judgment (i.e., P→J; I→J), while, in the second stage, perception and judgment influences decision choice (i.e., P→D; J→D) (see Figure 1.2).

The Throughput Model with its algorithmic pathways relates to machine learning, i.e., machine learning aided with the Throughput Model can contribute to the technological convergence of the growth of big data, decreasing data storage costs, increasing computing power, enhanced Artificial Intelligence algorithms and acceleration of cloud computing. Machine learning represents the ability for computers to learn without explicit programming.

Nonetheless, classic algorithms and machine learning are different. Algorithms are every so often designated as input-output machines. Traditional programming depends on functions that are embedded in logic—IF x, THEN y. Algorithms are rule based, explicit, and hard-wired. Machine learning is more intricate than traditional algorithms. Learning algorithms make their decision choices not by a pre-programmed proviso that their data must meet, but through the auditing and statistical analyses of hundreds or thousands of datasets in the realm that it makes the decision choice in the circumstances.

Understanding the Throughput Model algorithmic pathways is important since decision choices are related to reasoning in particular circumstances. One of the definitions of Artificial Intelligence refers to cognitive processes and especially to reasoning. Further, before making any decision, people also reason. Therefore, it is very informative to use a representation of the human mind such as the Throughput Model since it affords a natural link to Artificial Intelligence.

Further Examination and Impact of Artificial Intelligence and Algorithms

An algorithm is described as the process or the rules that problem-solving operations follow. Engineers use algorithms when designing a learning machine, the algorithm is what will make the machine process the data. Machine learning is the way to teach machines to adapt to changes within the technology or to adapt additional information to a current problem and make a rational decision. Machines have the advantage of speed, accuracy, and the amount of volume they can hold and still conclude, something that the human mind lacks.

Algorithms are a set of instructions depicting a preset, strict, coded recipe that gets implemented when it encounters a trigger. On the other hand, Artificial Intelligence is an exceedingly wide-ranging term covering a variety of Artificial Intelligence specializations and subsets. It includes a group of algorithms that can modify its algorithms and create new algorithms in response to learned inputs and data as contrasted to relying mainly on the inputs it was designed to recognize as triggers. This capability to modify, acclimatize and develop based on new data, is described as "intelligence".

Practitioners, researchers and commercial entities are taking advantage of Artificial Intelligence solutions with the implementation of enormous data footprints being generated in the process of daily activities through smart technologies such as personal digital assistants, location trackers, sensors, imaging devices and social media feeds.

Typically, Artificial Intelligence technology can be of benefit for the following reasons:

1. Cost of unhurried decisions is high (i.e., decision-making scenarios where speed is essential);
2. Cost of wrong decision choices is minimum;
3. Data size is too large for manual analysis or traditional algorithms;
4. Prediction accuracy is more important than explanation or clarification; and,
5. Regulatory requirements are slight.

In order to classify assorted categories of Artificial Intelligence, pertaining to implementation in organizations, we incorporate a framework developed by Kaplan and Haenlein, 2019, which was inspired by previous studies depicting effective supervisors and employees with superior performance (e.g., Boyatzis, 2008; Hopkins and Bilimoria, 2008; McClelland and Boyatzis, 1982; Stubbs Koman and Wolff, 2008). This literature typically accredits that higher performance is unmistakably related to the existence of three proficiencies or kinds of competencies: cognitive intelligence (e.g., competencies associated with pattern recognition and systematic thinking), emotional intelligence (e.g., adaptability, self-confidence, emotional self-awareness, and achievement orientation), and social intelligence (e.g., empathy, teamwork, and inspirational leadership) (Kaplan and Haenlein, 2019). Figure 1.3 delineates the different types of Artificial Intelligence tools, systems and softwares.

In sum, Artificial Intelligence use of machine learning (e.g., neural networks, deep learning) is a group of techniques for developing models that designate or predict something about an event. It does so by learning those models from data. Bayesian machine learning (especially for neural networks) provides the encoding of our prior beliefs regarding what those models should look like, independent of what the data tells us. Bayes' theorem is an approach to determine conditional probability. Conditional probability is the probability of an event occurring, given that it has some relationship to one or more other events.

In Artificial Intelligence, a new piece of evidence is bounded to the old evidence to configure the complete set of evidences. Bayes' rule stipulates how an entity should update its belief in a proposition based on a new piece of evidence. Also note that Bayesian neural network denotes covering standard networks with posterior inference. Standard neural network training by means of optimization is (from a probabilistic perspective) comparable to maximum likelihood estimation for the weights.

Given the popularity and importance of Artificial Intelligence in the global environment, machine learning provides the means to achieve an intelligent agent, which is powered by a diverse variety of programming languages nowadays such as python, tensorflow, java, and "C" (Table 1.2).

While the use of cognitive intelligence to classify Artificial Intelligence appears clearcut, the relevancy of emotional and social intelligence necessitates some enlightenment. The conventional viewpoint in psychology is that intelligence is generally innate (i.e., a feature that people are born with as opposed to something that can be learned). Notwithstanding, emotional and social intelligence are associated with specific emotional and social skills and it is these skills that individuals can learn and that Artificial Intelligence systems can replicate. While machines and Artificial Intelligence systems cannot evidently experience emotions themselves, they can be trained to recognize them (e.g., through analysis of facial micro-expressions) and consequently, modify their reactions.

In the face of its expanding presence across many facets of our lives, there is no commonly accepted definition of "artificial intelligence" (National Science and Technology Council, 2016). Instead, it is an umbrella term that consists of a variety of computational techniques and associated processes dedicated to cultivating the ability of machines to do things requiring intelligence, such as pattern recognition, computer vision, and language processing.

In this book, we have referred to Artificial Intelligence as machine intelligence or a machine's ability to replicate the cognitive functions of individuals. These tools have the capability to learn and solve problems. In computer science, these machines are fittingly described as "intelligent agents" or bots. In manufacturing, many significant benefits of robotics

Table 1.2. Components of Artificial Intelligence

TYPES OF MODELS	Machine learning, neural networks deep learning (using decision trees, forest trees, regressions, Bayes law, graphs).
PROGRAMMING LANGUAGES FOR MODEL BUILDING	Python, tensorflow, java, "C". These software programs take data and perform analytics, modeling and visualization.
SOFTWARE/HARDWARE FOR TRAINING AND RUNNING MODELS	GPUs, parallel processing tools (e.g., Throughput Model), cloud data storage platforms. Cloud computing involves delivering hosted services over the Internet. These services are broadly divided into three categories: Infrastructure-as-a-Service (IaaS), Platform-as-a-Service (PaaS) and Software-as-a-Service (SaaS).
APPLICATIONS	Image recognition, speech recognition, chatbots, natural language generation.

have improved the industry. That is, many of the manual and menial tasks are now performed by robots. They are controlled by Artificial Intelligence systems that direct how to cut, shape and combine materials. Robots can work 24 hours a day, and only take time off for maintenance and repair.

In addition, Artificial Intelligence is the study and creation of non-human apparatuses, such as computer systems that can perceive, reason and act. The intelligence can be exhibited by algorithms that imply pathways to solve problems (Rodgers and Al Fayi, 2019). Further, Artificial Intelligence has been depicted as "making a machine behave in ways that would be called intelligent if a human were so behaving" (McCarthy et al., 1955). Similarly, cognitive scientist Marvin Minsky deemed Artificial Intelligence as "the science of making machines do things that would require intelligence if done by men" (Minsky, 1968, p.v.).

Moreover, Artificial Intelligence is an interdisciplinary area embracing computer science, cognitive science, philosophy, neuroscience, linguistics, anthropology, etc., which exhibits the capacity of intelligence a computer system or robots have to be able to mimic what a human can do. Artificial Intelligence research also overlaps with tasks such as robotics, control systems, scheduling, data mining, logistics, speech recognition, facial recognition and many others.

The potential of Artificial Intelligence to enhance our lives is vast and multifaceted. For instance, Artificial Intelligence-based systems are already overtaking medical specialists in diagnosing certain illnesses, while the use of Artificial Intelligence in the financial system is increasing access to credit to borrowers that were once overlooked by decision makers. However, Artificial Intelligence also has shortcomings that stifle its huge potential. Artificial Intelligence-based systems infringe on privacy rights since they depend on the collection and usage of massive quantities of data to make predictions that in various cases, have served to propagate prevailing social patterns of bias and discrimination.

Nonetheless, because of the immense increase in accessible data assets, tumbling costs of computing and data storage, and substantial enhancements in system architecture and learning systems, Artificial Intelligence has made rapid strides in terms of its business applications. While many individuals fear the impact of Artificial Intelligence on employment, research reveals that organizations using Artificial Intelligence to propel innovation are more prone to enhance the size of their workforce rather than reduce it. If leveraged properly, Artificial Intelligence can play an essential role in improving profitability, amplifying efficiency, and unceasing innovation.

In view of this, there is a need for organizational leaders to become more accustomed with prevailing and developing technologies and their applications for business, as well as their inferences for leadership and management. The most successful supervisors of times ahead are those who can exploit our enlarged connectivity with machines to intensify organizational performance, and who distinguish that upskilling existing personnel to work effectively with technology is a tactical main concern.

One of the goals of Artificial Intelligence is to create robots or machines that are intelligent enough to perform the task of a human, without needing the human by its side at all times. Artificial Intelligence has the potential for machines to learn from experience, adjust to new inputs and perform human-like endeavors. For example, some of the Artificial Intelligence examples include chess-playing computers to self-driving cars, which relies greatly on components of machine learning such as deep learning and natural language processing (NPL).

Types of Expert Systems and Artificial Intelligence Systems

Expert systems are a collection of rules programmed by people in form of if-then statements. They are not a branch of Artificial Intelligence since they lack the capability to learn autonomously from external data. Moreover, expert systems embody a distinct approach altogether since they assume that people intelligence can be formalized through rules and hence reconstructed in a top-down approach (also called symbolic or knowledge-based approach).

If an expert system were programmed to recognize a human face, then it would check for a list of criteria (e.g., the presence of certain shapes, of a nose, of two eyes) before making an evaluation grounded on entrenched rules. Such systems tend to perform disappointedly during tasks that are subject to complex configurations of human intelligence, which are tacit and cannot be passed on without difficulty as simple rules. This is not to say that expert systems are not worthwhile and practical for everyday use. Moreover, IBM's famous Deep Blue chess-playing algorithm that defeated Garry Kasparov in the late 1990s, was not an Artificial Intelligence system, but an expert system. Expert systems like Deep Blue have been essential drivers in making Artificial Intelligence more pronounced in the global community.

There are fundamentally two distinct types of Artificial Intelligence, one is software-based and the other is physical agents, such as robots. Robots mimic what a person does, therefore performing the job of an individual, only requiring supervision. Physical agents were invented to be able to do day-to-day functions that a person is capable of performing; therefore, making it hassle-free for the tasks the individual has to perform. Software-based Artificial Intelligence should be able to do things that individuals cannot perform very fast. Examples include sorting through large amounts of data within seconds and making a conclusion based on their findings. Artificial Intelligence has been around for a long period; however, it has become more popular recently due to the amplification of technological advances and the increased amounts of data. In addition, Artificial Intelligence has many subsets including machine learning, neural networks and deep learning.

Arguably Artificial Intelligence as characterized above implements a bottom-up approach (also referred to as connectionist or behavior-based approach) by imitating the brain's structure (e.g., through neural networks) and exploiting large amounts of data to derive knowledge autonomously. This is similar to how children would learn to recognize a face—not by applying rules formalized by their parents but by seeing hundreds of thousands of faces and, at some point, being able to recognize what is a face and what is a chair. This permits dealing with chores and responsibilities immeasurably more complicated than what could be managed through expert systems. For instance, while chess can be formalized through rules, the Chinese board game Go is not constructed in this manner. Consequently, it was never probable to construct an expert system adept to defeat an individual Go player. Nevertheless, an Artificial Intelligence deep neural network can be trained to play Go simply by discerning a significant number of games played by people.

Nonetheless, these systems are probabilistic and may not be accurate for specific individuals. For example, deep learning computer vision systems can catalog an image similar to a person; but they will occasionally mis-specify a photo of a frog for a rifle (Conner-Simons, 2017). Further, chatbots are already becoming part of big business. Chatbots are implemented to replace automated telephone systems.

Finally, Artificial Intelligence driven cyberspace is not in one location, but in many places (Lessig, 2006). Moreover, the character of these many places differs in the ways that machine learning, neural networks and deep learning are implemented for use and application. Algorithms and decision trees help to support these tools, by incorporating codes that influence peoples' lives affected by these mechanisms.

To sum up, expert systems (i.e., software and physical agents) can be considered a *knowledge-based system* that are devoted to the idea of producing behavior by way of deduction from a set of axioms (Cristianini, 2014; Housel and Rodgers, 1994). These systems are "closed-rule algorithms". These "expert systems" implement formal logic and coded rules in reasoning. Examples are trust, real estate purchasing, tax preparation software, and healthcare diagnostic decision support algorithms. Further, these systems embody concrete circumstances as well as reasoning optimal decisions predicted on defined rules within a specific domain (Raso et al., 2018). Nonetheless, they cannot learn or automatically leverage the information they have amassed over time to enhance the quality of their decision-making (Buchanan, 1989).

Artificial Intelligence system involves technology which uses statistical learning in order to continuously enrich their decision-making performance. These tools include procedures such as machine learning, neural networks and deep learning, which is made possible by the exponential growth of computer processing power, decline in digital storage cost, and heightened data collection efforts (Tecuci, 2012). Systems in this classification consist of self-driving automobiles, facial and iris recognition systems utilized in monitoring, controlling and supervising. In addition, these systems include natural language processing tools that are implemented to automate translation and content moderation as well as algorithms that suggest what to view next on video streaming services (Raso et al., 2018).

The next section examines how the mechanisms of machine learning, neural networks and deep learning in Artificial Intelligence are embedded by certain features. These features can also change, and if they do, the values of Artificial Intelligence driven cyberspace will be different.

Machine learning: Basic concepts

Machine learning is a category of Artificial Intelligence that supplies computers and other machines with the ability to draw automatic inferences when exposed to new data, without being explicitly programmed. Further, this procedure uses certain techniques to bring about machines to perform certain tasks. Therefore, machine learning can be depicted as the knowledge of inducing computers to learn and act similar to people as well as enhancing their learning over time in an autonomous manner, by supplying it data and information in the form of observations and real-world interactions.

There are numerous examples of the current and future use of machine learning algorithms in decision-making, including (Royal Academy of Engineering, 2017; UK Government Office for Science, 2016):

1. In financial services industries, machine learning is being utilized to mechanize trading decisions and detect investment opportunities for clients.

2. In legal sectors, machine learning is employed to deliver legal advice to individuals and small businesses. In future, machine learning may allow for more informed decision-making by fashioning new insights from legal data and enabling better interaction with clients through new services.

3. In criminal justice systems, machine learning has been implemented to determine bail and prison sentences. In future, machine learning may allow for crime pattern detection, and for predictive policing.

4. In the education arena, machine learning is being utilized to rate teaching performance in schools. In future, machine learning could be used to improve learning efficiency by selecting assessments and other learning resources for each student individually.

5. In healthcare, machine learning is utilized to enhance the accuracy of diagnostics through pattern detection. In future, machine learning could be implemented to forecast responses to certain treatment pathways, permitting more informed decision-making around personalized treatment options.

6. Governments are presently implementing machine learning to provide insights into a wide range of data, which could be used in future to make informed decisions on existing services—such as healthcare or emergency services. Machine learning can also be utilized to inform the development of public policy in the future.

There are many varieties of machine learning algorithms, with hundreds published consistently overtime. Moreover, they are fundamentally assembled by either learning style (i.e., supervised learning, unsupervised learning, and semi-supervised learning) or by similarity in form or function (i.e., classification, regression, decision tree, clustering, deep learning, etc.). Notwithstanding learning style or function, according to Pedros Domingos, all combinations of machine learning algorithms comprise the following (see Table 1.3) (http://www. astro.caltech.edu/~george/ay122/cacm12.pdf):

Table 1.3. The three components of learning algorithms

Representation	Evaluation	Optimization
Instances	Accuracy/Error rate	Combinatorial optimization
K-nearest neighbor	Precision and recall	Greedy search
Support vector machines	Squared error	Beam search
Hyperplanes	Likelihood	Branch-and-bound
Naïve Bayes	Posterior probability	Continuous optimization
Logistic regression	Information gain	Unconstrained
Decision trees	K-L divergence	Gradient descent
Sets of rules	Cost/Utility	Conjugate gradient
Propositional rules	Margin	Quasi-Newton methods
Logic programs		Constrained
Neural networks		Linear programming
Graphical models		Quadratic programming
Bayesian networks		
Conditional random fields		

1. Representation (a set of classifiers or the language that a computer understands);
2. Evaluation (i.e., objective/scoring function); and
3. Optimization (search method; often the highest-scoring classifier: for example, there are both off-the-shelf and custom optimization methods used) (see Table 1.1).

How do we get machines to learn?

There are a variety of methods for inducing machines to learn, from using basic decision trees to clustering to layers of artificial neural networks (the latter of which has contributed to deep learning), subject to the task a person or organization is attempting to achieve. In addition, the type and amount of data available also conditions the method implemented to make machines to learn. This dynamics sees itself played out in applications as varying as medical diagnostics or self-driving automobiles.

Prominence is often placed on selecting the paramount learning algorithm. Nonetheless, researchers have found that some of the most interesting open problems in machine learning are those that arise during its application to real-world problems (Brodley et al., 2012). Research performed while functioning on real applications often drives progress in the field, and the explanations are twofold: (1) inclination to discover boundaries and limitations of prevailing methods; and (2) researchers and developers working with domain experts and putting their time and expertise to enhance system performance.

Machines that learn are instrumental to people since, with their huge processing power, they are able to more swiftly underline or find patterns in big (or other) data that would have been avoided or failed to have attended to by individuals. Machine learning is a tool that can be implemented to augment people's capability to decipher problems and make cognizant deductions on a broader assortment of problems, from assisting to diagnosing illnesses to coming up with solutions for green house effects.

The fundamental goal of machine learning algorithms is to take a broad view beyond the training samples. In other words, to interpret efficaciously data that it has never 'seen' before. As underscored in our definition of Artificial Intelligence earlier, a defining element of all those systems is the capability to learn from past data. For this, there are three broad types of machine learning processes: *unsupervised, supervised* and *semi-supervised*.

In *unsupervised learning process*, the inputs are labeled but not the outputs. This implies that the algorithm needs to deduce the underlying structure from the data itself. For example, cluster analysis targets at grouping elements in similar categories. In addition, where neither the structure of those clusters nor their number is known in advance, falls into this group. Since the output is derived by the algorithm itself, it is not possible to assess the accuracy or exactness of the resulting output. Therefore, users are required to place greater trust and confidence into the Artificial Intelligence system which can make managers uncomfortable. For instance, speech recognition made familiar with Apple's Siri or Amazon's Alexa can be conducted using unsupervised learning.

What is more, unsupervised machine learning is akin to thinking quickly on your feet, by gathering the information available and making a prediction based on that data. For instance, unsupervised learning is when you simply have input data (X) and no corresponding output variables. The goal of unsupervised learning is to model the fundamental structure or distribution in the data in order to discover more about the data.

Unsupervised learning occurs since there is no correct answers and there is no teacher. Algorithms perform on its own to in order to discover and present the noteworthy structure in the data.

Unsupervised learning problems can be further grouped into clustering and association problems.

- *Clustering*: A clustering problem exists in order to discover the inherent groupings in the data, such as grouping customers by purchasing behavior.
- *Association*: An association rule-learning problem exists to discover rules that describe large portions of your data, such as people who buy X also tend to buy Y.

Some prevalent examples of unsupervised learning algorithms are:

- k-means for clustering problems.
- *Apriori* algorithm for association rule learning problems.

Supervised learning methods map a given set of inputs to a given set of (labeled) outputs. They are typically desirable for managers because supervised learning includes methods many may be familiar with. For example, basic statistics, which include areas such as linear regression or classification trees. The more complex methods like neural networks also fall into this group. An example of supervised learning is to use a large database of labeled images to separate between those images showing a "cat" and those showing a "cake."

Supervised machine learning fashions the machine to make predictions based on functions to get from known inputs to known outputs. In other words, supervised learning is where you have input variables (x) and an output variable (Y) and you use an algorithm to learn the mapping function from the input to the output.

$$Y = f(X)$$

The objective is to estimate the mapping function so well that when you have new input data (x) that you can predict the output variables (Y) for that data.

This method is referred to as supervised learning since the process of an algorithm learning from the training dataset can be thought of as a teacher supervising the learning process. We know the correct answers; the algorithm iteratively produces predictions on the training data and is corrected by the teacher. Learning finishes when the algorithm accomplishes a satisfactory level of performance.

Moreover, supervised learning problems can be divided into regression and classification problems.

(1) *Classification*: A classification problem exists when the output variable is a category, such as "red" or "blue" or "disease" and "no disease".

(2) *Regression*: A regression problem occurs when the output variable is a real value, such as "dollars" or "weight."

Some typical kinds of problems constructed on top of classification and regression include recommendation and time series prediction respectively.

Some widespread examples of supervised machine learning algorithms are:

(a) Linear regression for regression problems.
(b) Random forest for classification and regression problems.
(c) Support vector machines for classification problems.

Problems that exist, which have a large amount of input data (X) and only some of the data is labeled (Y) are called *semi-supervised learning* problems. These problems are situated between both supervised and unsupervised learning. A good example is a photo archive where only some of the images are labeled (e.g., person, cat, dog), and the majority are unlabeled.

Image recognition capabilities can also be used to identify whether the same official appears in multiple images or if officials from a location other than the intended site have uploaded photos. Considering the scale of this initiative, which involves creating more functional toilets, being able to check every image rather than a small sample will actually help increase effectiveness.

Many real-world machine learning problems fall into this area. This is because it can be expensive or time-consuming to label data as it may require access to domain experts. Whereas unlabeled data is cheap and easy to collect and store.

Further, in semi-supervised learning, the system receives an output variable to be maximized and a series of decisions that can be taken to impact the output. For example, Artificial Intelligence system that aims to learn playing Pac-Man, simply by knowing that Pac-Man can move up, down, left and right and that the objective is to maximize the score attained in the game. Software giant Microsoft uses semi-supervised learning to select headlines on MSN.com by rewarding the system with a higher score when more visitors click on a given link.

The use of unsupervised learning procedures can uncover and learn the structure in the input variables. Further, the implementation of supervised learning procedures can be operationalized to make best guess predictions for the unlabeled data, feed that data back into the supervised learning algorithm as training data and use the model to make predictions on new unseen data.

Neural networks basic concepts

The very first artificial neural network was designed by Minsky as a graduate student in 1951 (see McCarthy et al., 1955); however, the approach was limited at first. Hecht-Nielsen (1988) states that a neural network as a computing system, which consists of a number of simple, highly interconnected processing elements. These elements process information by their dynamic state response to external inputs.

The structure of artificial neural networks was based on the present understanding of biological neural systems. Therefore, artificial neural networks are a limited abstraction of the human brain; thus, the organization of the artificial neural system is very analogous to the one of biological neurons. The all-embracing understanding of biological neurons is not complete; nonetheless, the main functionality that adds to the learning capability of a system is employed in artificial neural networks. The essential element, an artificial neuron, is a model based on known behavior of biological neurons that demonstrate most of the characteristics of the human brain.

This model computation is achieved by dense interconnection of simple processing units. To describe the attributes of computing, the artificial neural networks go by many names such as connectionist models, parallel distributed processors, or self-organizing system. With such features, an artificial neural system has great potential in performing applications such as speech and image recognition where intense computation can be done in parallel and the computational elements are connected by weighted links.

In sum, the artificial neuron, the most elementary computational unit, is modeled based on the basic property of a biological neuron. This type of processing unit performs in two stages: weighted summation and some type of nonlinear function. It accepts a set of inputs to generate the

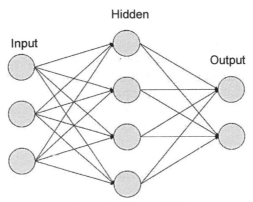

Figure 1.3. Neural network diagram

weighted sum, then passes the result to the nonlinear function to make an output (see Figure 1.3).

Since computer scientists first attempted the original artificial neural network, cognitive neuroscientists have learned a great deal regarding the human brain. One of the aspects they learned is that distinct parts of the brain are accountable for processing different facets of information and these parts are organized hierarchically. Therefore, input comes into the brain and each level of neurons offer insight and then the information is delivered on to the next higher level. That is exactly the mechanism that artificial neural networks (ANNs) are attempting to replicate. ANNs can be also described as parallel distributed computing models (Rodgers, 1991b). The fundamental processing units, neurons, are highly connected with strengths that are vigorously transformed during the system's learning process.

Artificial neural networks use different layers of mathematical processing to make sense of the information it is provided for use. By and large, an artificial neural network has somewhere from dozens to millions of artificial neurons—called units—assembled in a series of layers. The input layer receives numerous kinds of information from the outside world. This data goes into the network, which directs to process or learn about the features or patterns. From the input unit, the data moves through one or more hidden units. The hidden unit's chore is to change the input into something the output unit can implement.

The multitude of neural networks are effusively connected from one layer to another. These associations are weighted; the higher the number the more influence one unit has on another, comparable to a human brain. As the data goes through each unit, the network is learning additionally about the data. The other network part is the output units. This part of the network reacts to the data that was provided and processed in the system.

Unlike conventional computing systems, which has fixed instructions to perform specific computations, ANNs need to be taught and trained to function correctly. The advantage is that the neural system can learn new input-output patterns and adjust the system parameters. Such learning can eliminate specifying instructions to be executed for computations. Instead, users simply supply appropriate sample input-output patterns to the network. The model of the entire ANNs are determined by the network topology, type of neural model, and learning rules. These are the main interests in designing ANNs.

Characteristics of Artificial Neural Networks (ANNs)

The ANNs are representative of a large number of very simple processing neuron-like processing elements that can process information in parallel with high speed and in a distributed manner (Rodgers, 1991b). A large number of weighted connections between the elements distributed representation of knowledge over the connections that knowledge is acquired by network through a learning process robustness and fault tolerance: The decay of nerve cells does not seem to affect the performance significantly. These systems are latent with flexibility, capability and collective computation.

Flexibility is that the network can automatically adjust to a new environment without using any programmed instruction. Next, capability is the process to deal with a variety of data situations, which indicates that the network can deal with information that is fuzzy, probabilistic, noisy and inconsistent. Finally, *collective computation* refers to the network performing routinely many operations in parallel as well as given tasks in a distributed manner. Therefore, the ANNs display some brain-like behaviors that are difficult to program.

Directly such as:

(a) Learning,
(b) Association,
(c) Categorization,
(d) Generalization,
(e) Feature extraction,
(f) Optimization, and
(g) Noise immunity.

Deep learning basic concepts (including natural language processing)

Deep learning can be depicted as a newer area of machine learning that utilizes multi-layered artificial neural networks to deliver high accuracy in chores such as object detection, speech recognition, language translation, etc. Deep learning is a collection of procedures that will study what the human brain does for a computer then to reconstruct. Deep learning refers to the use of artificial neural networks (ANNs) in order to smooth the progress of learning at multiple layers. It is a part of machine learning based on how data is presented, instead of task-based algorithms. Further, deep learning models are trained by using large sets of data that learn features directly from the data without the requirement for manual feature extraction. Deep learning is also distinct from machine learning in that it doesn't require human operators to command the machine or translate its output. Deep learning is mostly unsupervised and aims to avoid the need for human intervention.

In recent years, a fair number of scientists and innovators began to devote their work to Artificial Intelligence systems. Moreover, technology has finally caught up with faster and more powerful general processing units (GPUs). Industry observers staple this resurgence to 2015, when fast and powerful parallel processing became at hand. In addition, this is also around the birth of the so-called Big Data movement, when it became feasible to store and analyze infinite amounts of data.

Deep learning has led the way in transforming analytics and facilitating practical applications of AI. Deep learning examples includes automatic photo-tagging on Facebook. In addition, Blipparhas developed an augmented reality application that put to use deep learning in real-time object recognition in 2015 (https://www.tenfold.com/business/artificial-intelligence-business-decisions).

Accordingly, we in the era of deep learning. Again, deep learning refers to the use of ANNs in order to smooth the progress of learning at multiple layers. It is a part of machine learning based on how data is presented, an alternative to task-based algorithms. Moreover, deep learning is based mostly on neural networks whereby each level learns to transform its input data into a slightly more abstract and composite representation. Whereas, natural language processing assists computers to understand, interpret and manipulate human language. Natural language processing advances from different disciplines, including computer science and computational linguistics, in its quest to fill the gap between human communication and computer understanding. Deep learning has led the way in revolutionizing analytics and facilitating practical applications of Artificial Intelligence (Table 1.4).

Natural language processing

Natural language processing (NLP) is the capability of a computer program to recognize human language as it is spoken. NLP is a component of artificial intelligence (AI). Hence, NLP allow computers to analyze, understand,

Table 1.4. Artificial Intelligence—Machine learning, neural networks & deep learning

Artificial Intelligence	Machine Learning	Neural Networks	Deep Learning
A feature of machines that symbolize a method of intelligence, rather than merely transmitting computations that are input by individual users.	A tactic to Artificial Intelligence whereby an algorithm learns to make predictions from data that is supplied into the system.	A machine learning approach whereby algorithms process signals by means of interconnected nodes described as artificial neurons.	A type of machine learning that often utilizes a network with several layers of computation (i.e., a deep neural network) facilitating an algorithm to strongly evaluate the input data.
Early purposes of Artificial Intelligence encompassed machines that could participate in games such as checkers and chess and programs that could scrutinize and replicate language.	From personalized news feeds to weather prediction maps, most individuals in developed countries use machine learning-based technologies every day.	Since they simulate the architecture of biological nervous systems, artificial neural networks (ANNs) are the recognizable technique of choice for modeling the brain.	Deep neural networks are responsible for self-driving vehicles, which learn to recognize traffic signs, as well as for voice-controlled virtual assistants.

and derive meaning from human language in a smart and useful way. Moreover, NLP considers the hierarchical structure of language: several words make a phrase, several phrases make a sentence and, ultimately, sentences convey ideas. By utilizing NLP, developers can organize and structure knowledge to perform tasks such as automatic summarization, translation, named entity recognition, relationship extraction, sentiment analysis, speech recognition, and topic segmentation.

The field of NLP encompasses a wide variety of topics which involve the computational processing and understanding of human languages. Beginning in the 1930s (Hutchins, 2001), the field historically has focused on the use of human-crafted rules, ad-hoc processing, and mathematical logic. Further, natural language processing gained momentum when Roberto Busa, an Italian Jesuit priest was a pioneer of computational linguistics. He evaluated the comprehensive compositions of St. Thomas Aquinas, the 13th-century Catholic priest and philosopher. The first ANNs were conceived in the 1940s as computational devices to capture human intelligence. For example, in 1949, Busa met with IBM founder Thomas J. Watson and convinced him to promote the *Index Thomisticus*, a computer-readable compilation of Aquinas' works. This project took more than 30 years and in due course was published in 56 volumes centered on more than 11 million computer punch cards, one for every word analyzed (Winter, 1999).

Since the 1980s, NLP has gradually depended upon data-driven computation involving statistics, probability, and machine learning (Jones, 1994; Liddy, 2001). The use of ANNs and deep learning has upsurge considerably in the field (Goldberg, 2017; Liu and Zhang, 2018). That is, recent enhancements in computational power and parallelization, exploited by Graphical Processing Units (Coates et al., 2013), now allow for "deep learning" that utilizes ANNs, sometimes with billions of trainable parameters (Goodfellow et al., 2016). In addition, the contemporary accessibility of large datasets assisted by sophisticated data collection processes, made possible the training of such deep architectures by means of their associated learning algorithms (LeCun et al., 2015; Schmidhuber, 2015).

Current approaches to NLP are based on deep learning, which analyzes and utilizes patterns in data to enhance a program's understanding. Deep learning models have a need to large amounts of labeled data to train on and identify relevant correlations. At the present time, assembling this kind of "big data" set is one of the main impediments to NLP.

How natural language processing operates

Syntax and semantic analysis are two primary tools utilized with natural language processing. Syntax is the assembly of words in a sentence that affords grammatical understanding to people. NLP utilizes syntax to evaluate meaning from a language grounded on grammatical rules. Syntax techniques implemented contain *parsing* (grammatical scrutiny for a sentence), *word segmentation* (i.e., partitions a large part of text to units), *sentence breaking* (i.e., places sentence boundaries in large texts), *morphological segmentation* (i.e., break up words into groups) and *stemming* (i.e., separates words with inflection in them to root forms).

Semantics encompasses the utilization and meaning related to words. NLP directs algorithms to understand the meaning and structure of sentences. Techniques that NLP implements with semantics consist of word sense disambiguation (i.e., derives meaning of a word based on context), named entity recognition (i.e., determines words that can be categorized into groups), and natural language generation (i.e., uses a database to determine semantics behind words).

Nevertheless, the expansion of NLP functions is challenging since computers customarily necessitate individuals to "speak" to them in a programming language that is precise, explicit and decidedly structured. Or these functions represent a limited number of clearly enunciated voice commands. Nonetheless, human speech is not at all times precise. That is, it is often equivocal, and the linguistic structure can be governed by several complex variables, consisting of slang, regional dialects and social context.

Uses of natural language processing

Research conducted on NLP revolves around search, especially enterprise search. Enterprise search is the organized retrieval of structured and unstructured data within an organization (Kruschwitz and Hull, 2017). Further, produces content from multiple enterprise-type sources, such as databases and intranets, searchable to a defined audience. Utilized appropriately, enterprise search generates an easily navigated interface for entering, categorizing and retrieving data securely, in compliance with security and data retention regulations.

This includes permitting users to query data sets in the form of a question that they might present to another individual. The machine deuces the essential components of the human language sentence, such as those that might match up to specific characteristics in a data set and returns an answer.

NLP can be utilized to decipher free text and make it evaluable. There is a incredible amount of information stored in free text files, such as patients' medical records. Before deep learning-based NLP models, this information was inaccessible to computer-assisted analysis and could not be analyzed in any systematic way. Nonetheless, NLP permits users to filter through vast amounts of free text in order to locate relevant information in the files.

Some of the capabilities that NLP-based systems can provide to organizations include:

- *NLP (Categorization)*: NLP can be utilize to categorize text, create summaries of documents, permitting indexing and searching for duplicate content in large volumes of text or speech data.

- *NLP (Topic Identification)*: NLP can potentially ascertain the meaning and bigger topics within considerable volumes of text data.

- *NLP (Extraction and Abstraction)*: NLP can be implemented to retrieve relevant paragraphs from a prolong text document or develop a summarized version of the document that contains only the most relevant paragraphs.

- *NLP (Information Retrieval)*: NLP can be utilized in order to search for specific information within digital text.

- *NLP (Intent Parsing)*: NLP can discriminate the context within human language. Essentially, chatbots require NLP for intent parsing since it deals with recognizing the intent within text data. Further, it can respond to customers with text. In total, the NLP software necessitates to "learn" the applicable text responses to text it receives.

- *NLP (Sentiment Analysis)*: NLP can establish the tone and inferred opinions (positive or negative) behind human language in text form.

- *NLP (Speech/Voice Recognition)*: NLP can convert natural language human speech into written text, and vice versa.
- *NLP (Machine Translation)*: NLP can be implemented to robotically translate text or speech data from one language to another.

Taken as a whole, the aim throughout these functions is to take basic human speech or text data to make it invaluable. This is performed by using Artificial Intelligence and its subset of machine learning to extricate insights or add value to the data.

Using these technologies, computers can be trained to accomplish specific tasks by processing large amounts of data and recognizing patterns in the data. Artificial Intelligence also denotes to the capability of a computer or a computer-enabled robotic system to process information and produce outcomes in a method comparable to the thought process of people in learning, decision-making and solving problems. By extension, the goal of Artificial Intelligence systems is to develop systems proficient in handling complex problems in ways analogous to human logic and reasoning.

In deep learning procedures, there is no need teach the computer how to perceive the information or where to apply it, the computer recognizes what to do with the information. Deep learning goes an additional level than machine learning being able to interpret new or unknown information and going deeper into additional layers of information. Some good examples of deep learning relate to something as small as cellular devices to something as large as self-driving cars. As technology progresses the deep learning in machines becomes more superior, in cell phone while on a search-engine many advertisements will pop-up and those advertisements will be something further to investigate. A cell phone can commit to memory what was previously searched and store that information for later. In addition, when the search engine opens again it will try to go back to what was examined previously. Self-driving cars also became very popular when Tesla, a car manufacturer, was the first company to create a car that did not require a driver. The self-driving car will be able to take a person to and from places safely just as a standard automobile can go to a particular destination. The technology in this kind of automobile recognized all the traffic signals, such as stop signs and traffic lights just as any individual can follow. The concept of self-driving cars is still improving; however, as more information becomes available and technology advances it will be become proficient at a high level.

In sum, deep learning is an advancement over other AI tools at inducing computers to handle a variety of skills, such as understanding photos. Deep learning software can even understand sentences and respond with appropriate answers, make questions clearer or provide suggestions of its own.

Conclusion

Since the invention of machines and computers, its capability to execute many tasks has grown exponentially. Individuals have developed computer systems by cultivating their power in terms of sundry working domains, amplifying their speed and shrinking their size over time. The following Chapter 2 discusses how the Throughput Model (see Figure 1.2) depicts information processed by people thinking processes or Artificial Intelligence by combining machine learning, neural networks, algorithms, and deep learning. The Throughput Model is partially based upon a conceptual framework of decision-making from previous studies of Rodgers (1984, 1991a,b, 1997, 2006, 2016). These studies suggested that decision makers rely upon two stages to form decision choices. The first stage conceptualized individuals' framing of a problem (perception) and their covariation with information as well as influencing judgment (i.e., P→J; I→J). Further, this model also indicated that decision makers' perceptions act interdependently with certain information cues to influence and guide subsequent processes of information in the second stage (i.e., P→D; J→D).

Artificial Intelligence depicts human intelligence exhibited by a machine. It is the ability of machines to inculcate the ability to think, to build perceptions, plan in advance and to respond like people. One of the many aims of Artificial Intelligence is to develop problem solving or decision-making skills in order to assist people with their daily chores efficiently and effectively. Machine learning is a subset of Artificial Intelligence since the decisive goal is to make machines self-dependent, whereby it can make decisions based on logic, reasoning, rational thinking and past experiences.

For machine learning the difference among unsupervised, supervised and semi-supervised learning. You now know that:

(a) *Unsupervised*: All data is uncategorized, and the algorithms learn to inherent structure from the input data.

(b) *Supervised*: All data is categorized, and the algorithms learn to forecast the output from the input data.

(c) *Semi-supervised*: Some data is categorized but most of it is uncategorized and an assortment of supervised and unsupervised techniques can be implemented.

The differences between machine learning and Artificial Intelligence is that the former is about learning through data and past experiences and taking decisions on its basis. Whereas, Artificial Intelligence is more global in that it pertains to the ability of the machines to interact with individuals and to assist them in their day to day activities.

Artificial Intelligence systems can be very robust and are quickly on the increase. They make available outputs that can be exceptionally accurate, replacing and, in some cases, far supplanting individuals' efforts. Nevertheless, these systems do not reproduce human intelligence. The global society should distinguish the strengths and limits of this distinct form of intelligence, as well as construct an understanding of the preeminent ways for individuals and computers to work together.

In our modern-day society, AI-based functions have already touched people's lives in an assortment of numerous aspects. From the intelligent keyboards on smartphones to the voice-activated assistants in tablets and desktops and the devices in a person's proximate personal space; technology has become increasingly far more intelligent than it was in yesteryear or is perceived to be by people or organizations. Whether it is financial services, healthcare, education or even security and governance, Artificial Intelligence can be utilized for the benefit of everyone. AI-based automation is capable of shaping almost every sector of the global economy and society.

Artificial Intelligence is an interdisciplinary field that necessitates knowledge in computer science, linguistics, psychology, biology, philosophy, etc., for research. Artificial Intelligence can also be defined as the area of computer science, which deals with the approaches that computers can be made to perform cognitive functions attributed to people. Moreover, Artificial Intelligence draws from the fields of computer science, cognitive science, engineering, ethics, linguistics, logic, mathematics, natural sciences, philosophy, psychology, and statistics.

Deep learning has led the way in revolutionizing analytics and enabling practical applications of AI. We see it in something as basic as automatic photo-tagging on Facebook.

We can look forward to driverless cars and so much more. In a seminar view, we can expect Artificial Intelligence to be applied further in profit and non-profit organizations, particularly in decision-making.

In the coming years, artificial intelligent systems will take over more and more decision-making tasks from individuals. The worldwide scientific society has journeyed a distance since the development of Artificial Intelligence as a concept to its contemporary attraction as a field with near-boundless potential in turning around the manner in which activities are achieved in an operational society. The foremost frontier for Artificial Intelligence systems continues to be accomplishing a level of complexity that corresponds to the human mind.

The recent growth in the Artificial Intelligence research and practice arena, namely on the data intensive sub-symbolic side of the AI technologies portfolio has implications for the information computer technology hardware related areas. Recognition of research in information computer technology hardware should keep up with the advances made in AI technologies and that better links between and across these communities are highlighted in the following chapters.

In the short to medium term, Artificial Intelligence brings many opportunities for decision-makers to improve their efficiency, provide more insight and deliver more value to businesses. In the longer term, AI raises opportunities for much more radical change, as systems increasingly take over decision-making tasks currently performed by individuals. This book outlines a framework for embracing the opportunities created by increasingly intelligent systems, based on three questions.

(a) What is the long-term vision for the global community?

We need to envision how intelligent systems can enable better decisions in business, and understand how accountants can help this process.

(b) How do artificial intelligence, decision-making, ethics and trust work together?

We need to develop an understanding of what is new about the technology, how it can 'turbo charge' the capabilities of individuals' trust and ethical positions, and its limits.

(c) How are decision-makers using Artificial Intelligence capabilities?

Artificial Intelligence technology combined with decision-making is not the end but only a means towards effectiveness and efficiency. Enhanced innovative capabilities, and better opportunities are already the result of Artificial Intelligence tools. Furthermore, several industries have begun to adopt Artificial Intelligence into their operations.

Most major industries are affected by AI. These industries employ "narrow AI", which performs objective functions using data-trained models and often falls into the categories of machine learning, neural networks, and deep learning. This is especially bona fide in the past several years, as data collection and analysis has geared up considerably thanks to robust IoT connectivity, the proliferation of connected devices and ever-speedier computer processing.

Some sectors are at the start of their Artificial Intelligence journey, others are veteran journeyers. Nonetheless, both have quite a distance to go. Notwithstanding, the impact that Artificial Intelligence is having globally is difficult to close the eyes to the following industries:

Transportation: Even though it may take some time to perfect autonomous vehicles, they will be increasingly a part of our society.

Manufacturing: Artificial Intelligence powered robots operate in conjunction with people in order to perform a constrained range of tasks such as assembly and stacking as well as predictive analysis sensors that monitor equipment to run efficiently.

Healthcare: In the comparatively Artificial Intelligence blossoming field of healthcare, diseases are more quickly and precisely identified, drug discovery time is reduced, virtual nursing assistants monitor patients, and big data analysis assist to improve a more personalized patient experience.

Education: With the assistance of AI, textbooks are digitized, early-stage virtual tutors assist individual teachers. Moreover, facial recognition measures students' emotions in order to determine who is struggling or bored; thereby, enhancing the experience to their individual needs.

Media: Artificial Intelligence is benefiting journalism. For example, Bloomberg implements Cyborg technology to aid effective logic of complex financial reports. Further, the Associated Press utilizes the natural language in order to produce earning report stories.

Customer Service: Google is employing an Artificial Intelligence assistant that can place people-like calls to make appointments at retail services such as hair salon. In addition to words, the system is familiar with context and nuance.

In sum, machine learning utilizes algorithms to break down data, learn from that data, and formulate informed decision choices grounded upon what it has learned. Deep learning assembles algorithms in layers to produce an artificial "neural network" that can learn and fashion intelligent decision choices on its own. Deep learning can be regarded as a subfield of machine learning. While both fall under the broad category of AI, deep learning is generally what is the inspiration for most human like AI.

In this book, we provide examples of individuals/organizations using Artificial Intelligence systems, including the specific benefits, ethical considerations and limitations, to help us develop the longer-term vision. Later chapters will explore how a strong AI community that is actively engaged in the challenges of responsible research and innovation and public acceptability of artificial intelligence, ensuring that research outcomes are socially beneficial, ethical, trusted, and deployable in real world situations.

References

Boyatzis, R.E. 2008. Competencies in the 21st century. Journal of Management Development, 27(1): 5–12.

Brodley, C.E., Rebbapragada, U., Small, K. and Wallace, B.C. 2012. Challenges and opportunities in applied machine learning. Artificial Intelligence Magazine, 33(1): https://doi.org/10.1609/aimag.v33i1.2367.

Buchanan, B.G. 1989. Can machine learning offer anything to expert systems? Machine Learning, 4(3-4): 251–54, https://doi.org/10.1023/A:1022646520981.

Coates, A., Huval, B., Wang, T., Wu, D., Catanzaro, B. and Andrew, N.G. 2013. Deep learning with COTS HPC systems. In International Conference on Machine Learning, 1337–1345.

Conner-Simons, A. 2017. Fooling Neural Networks w/3D-Printed Objects, MIT Computer Science & Artificial Intelligence Lab (blog), November 2nd, https://www.csail.mit.edu/news/fooling-neural-networks-w3d-printed-objects.

Cristianini, N. 2014. On the current paradigm in artificial intelligence. AI Communications 27(1) (January): 37–43, https://doi. org/10.3233/AIC-130582.

Domingos, P. 2019. A Few Useful Things to Know about Machine Learning. https://homes.cs.washington.edu/~pedrod/papers/cacm12.pdf.

Goodfellow, I., Bengio, Y., Courville, A. and Bengio, Y. 2016. Deep Learning. Vol. 1. Cambridge: MIT Press.

Hammond, K.R., McClelland, G.H. and Mumpower, J. 1980. Human Judgment and Decision Making: Theories, Methods, and Procedures. New York: Praeger.

Hecht-Nielsen, R. 1988. Theory of the backpropagation neural network. Neural Networks, 1(Supplement-1): 445–448.

Hopkins, M.M. and Bilimoria, D. 2008. Social and emotional competencies predicting success for male and female executives. Journal of Management Development, 27(1): 13–35.

Housel, T. and Rodgers, W. 1994. Cognitive styles reconceptualized: A multi-stage decision making model. International Journal of Intelligent Systems in Accounting, Finance and Management, 3(1994): 165–186.

Hutchins, W.J. 2001. Machine translation over fifty years. HistoireEpistémologieLangage 23(1): 7–31.

Jones, K.S. 1994. Natural language processing: a historical review. In Current Issues in Computational Linguistics: in Honor of Don Walker. London: Springer, 3–16.

Kaplan, A. and Haenlein, M. 2019. Siri, Siri, in my hand: Who's the fairest in the land? On the interpretations, illustrations, and implications of artificial intelligence. Business Horizons, 62: 15–25.

Kruschwitz, U. and Hull, C. 2017. Searching the enterprise. Foundations and Trends in Information Retrieval, 11: 1–142.

LeCun, Y., Bengio, Y. and Hinton, G. 2015. Deep learning. Nature, 521(7553): 436–444.

Lessig, L. 2006. CODE. New York: Perseus Books Group.

Liddy, E.D. 2001. Natural language processing (2001). In Encyclopedia of Library and Information Science, NY: Marcel Decker, Inc.

McCarthy, J., Minsky, M.L., Rochester, N. and Shannon, C.E. 1955. A proposal for the Dartmouth summer research project on artificial intelligence. Available at http://www-formal.stanford.edu/jmc/history/dartmouth/ dartmouth.html.

McClelland, D.C. and Boyatzis, R.E. 1982. Leadership motive pattern and long-term success in management. Journal of Applied Psychology, 67(6): 737–743.

Minsky, M.L. 1968. Semantic Information Processing. Cambridge, MA: MIT Press.

National Science and Technology Council: Committee on Technology. 2016. Preparing for the Future of Artificial Intelligence, Government Report (Washington, D.C.: Executive Office of the President, October).

Raso, F., Hilligoss, H., Krishnamurthy, V., Bavitz, C. and Kim, L. 2018. Artificial Intelligence & Human Rights: Opportunities & Risks. Berkman Klein Center for Internet & Society Research Publication Series. Research Publication No. 2018-6. https://cyber.harvard.edu/sites/default/files/2018-09/2018-09_AIHumanRightsSmall.pdf?

Rodgers, W. 1984. Usefulness of decision makers' cognitive processes in a covariance structural model using financial statement information. University of Southern California, dissertation.

Rodgers, W. 1991a. Evaluating accounting information with causal models: Classification of methods and implications for accounting research. Journal of Accounting Literature, 10: 151–180.

Rodgers, W. 1991b. How do loan officers make their decisions about credit risks? A study of parallel distributed processing. Journal of Economic Psychology, 12: 243–265.

Rodgers, W. 1992. The effects of accounting information on individuals' perceptual processes. Journal of Accounting, Auditing, and Finance, 7: 67–95.

Rodgers, W. 1997. Throughput modeling: Financial information used by decision makers, Bingley: Emerald Group Publishing.

Rodgers, W. 2006. Process Thinking: Six Pathways to Successful Decision Making. NY: iUniverse, Inc.

Rodgers, W. 2016. Knowledge Creation: Going Beyond Published Financial Information. Hauppauge, NY: Nova Publication.

Rodgers, W. and Al Fayi, S. 2019. Ethical Pathways of Internal Audit Reporting Lines, Accounting Forum 43(2).

Royal Academy of Engineering. 2017. Algorithms in decision-making A response to the House of Commons Science and Technology Committee inquiry into the use of algorithms in decision-making (April). https://www.raeng.org.uk/publications/responses/algorithms-in-decision-making.

Schmidhuber, J. 2015. Deep learning in neural networks: An overview. Neural Networks, 61: 85–117.

Stubbs, Koman, E. and Wolff, S.B. 2008. Emotional intelligence competencies in the team and team leader: A multi-level examination of the impact of emotional intelligence on team performance. Journal of Management Development, 27(1): 55–75.

Tecuci, G. 2012. Artificial Intelligence, Wiley Interdisciplinary Reviews: Computational Statistics 4, no. 2 (2012): 168–80, https://doi. org/10.1002/wics.200.

UK Government Office for Science. 2016. Artificial intelligence: opportunities and implications for the future of decision making, www.gov.uk/government/uploads/system/uploads/attachment_data/file/566075/gs-16-19-artificialintelligence-ai-report.pdf.

Winter, T.N. and Roberto Busa, S.J. 1999. The invention of the machine-generated concordance, University of Nebraska-Lincoln, Faculty publications, Classics and religious studies department (January).

2

Prelude to Artificial Intelligence
Decision-Making Techniques

"If I have seen further it is by standing on the shoulders of Giants."

—Isaac Newton

"Basically, there's no institution in the world that cannot be improved with machine learning."

—Amazon CEO Jeff Bezos

When we buy something on Amazon or watch something on Netflix, we think it's our own choice. Well, it turns out that algorithms influence one-third of our decisions on Amazon and more than 80% on Netflix. What's more, algorithms have their own biases. They can even go rogue.

—Kartik Hosanagar, University of Pennsylvania, Wharton Business School, USA

The vision for Artificial Intelligence should be piloted by innovative thinking, which includes the long-term objective of enhanced, or new strategic models. Previous to the unveiling of Artificial Intelligence, organizations had to rely on inconsistent data. Therefore, the decision-making process was not very precise. Now, with Artificial Intelligence, it is possible for organizations to turn to simulations, data-based and decision-making models. In addition, updated Artificial Intelligence systems begin from zero and feed themselves with a regular stream of Big Data. This is intelligence in action and eventually provides sophisticated data models that can be implemented for precise decision-making models.

Now that individuals have programmed computers to learn, the Throughput Model's algorithmic pathways can assist users in examining how Artificial Intelligence-supported software are learned and how they make decisions after their learning process is complete. The answers to such questions may provide additional light on our own decision-making processes, i.e., Throughput Modelling provides for the combining of human decision making with Artificial Intelligence, information technology, and systems engineering.

Moreover, the Throughput Model is a first step to evaluate explainable models for Artificial Intelligence decision-making processes. This process is useful and informative regarding the explanation that depicts the underlying processes of perception and judgment. The application of the Throughput Model has been applied to the areas of auditing, commercial lending, ethics, finance, military decision-making, trust modeling, etc., in part, due to what the model can deliver regarding neural networks and algorithmic pathways (Rodgers, 1999, 2012). At the same time, Artificial Intelligence techniques can be applied to the Throughput Model in order to provide computational assistance to humans in practical applications. In this chapter, the discussion centers on the Throughput Model's algorithmic pathways along with ways Artificial Intelligence can assist in the decision-making process for organizations.

Further, the Throughput Model is presented in order to enhance a more informed, strategic decision-making process, and augment organizational performance by integrating key Artificial Intelligence methods into the way an organization operates. As discussed in the previous chapter, Artificial Intelligence refers to machine intelligence or a machine's ability to imitate the cognitive functions of an individual. Artificial Intelligence tools and methodologies have the capability to learn and solve problems. In computer science, these machines are fittingly described as "intelligent agents" or bots. This chapter covers the areas of: (1) Artificial Intelligence and decision-making, (2) advantages of Artificial Intelligence, (3) Throughput Modeling algorithms, (4) human biases and decision-making, and (5) an example dealing with algorithms for decision-making processes for loan sanctioning officers.

Artificial Intelligence and Decision-Making

Artificial Intelligence tools are everyday expanding and enhancing decision-making. This technology not only provides for coordinating diverse data sources delivery in an effective and efficient manner, but also analyzes evolving data sources, trends, provides defined forecasts, develops data consistency, and quantifies uncertainty of all data variables. In addition, Artificial Intelligence tools attempt to anticipate the individual or machine user's data requirements, as well as provide information to a person or machine user in the most appropriate forms. Finally, Artificial Intelligence tools suggest critical courses of all possible action based on the intelligence gathered. Artificial Intelligence tools are necessary in a globally fast-changing digital age environment, since it is very arduous for individual decision-makers to keep up, analyze the mounds of data in front of them, and make informed and intelligent decisions.

Thus, one of the many areas that Artificial Intelligence covers is cognitive processes, particularly relating to the reasoning process. Before making any decision choice, people also reason; therefore, it is natural to explore the links between Artificial Intelligence and decision-making.

Artificial Intelligence based decision-making can provide useful advantages to organizations such as:

1. *Speed of decision-making*: computing power can produce optimal decisions far quicker than individuals.

2. *Minimizing human bias and error*: algorithms are not influenced by emotion and impulsive decision-making.

3. *Increasing coverage of data considered*: cross-analysis of data from markets, geo-political, economic, currency headwinds, traditional and social media, amongst many others.

4. Appraising multiple "what if" scenarios.

5. *Make the most on heretofore unheard-of market opportunities*: coming into new markets through technological disruption not thought of within the original human field of thought or vertical integration strategies.

Artificial Intelligence is becoming more prevalent in many aspects of decision-making in the present as well as the future. The application of algorithms is exponentially expanding as vast amounts of data are being created, captured and analyzed by governments, businesses and public bodies. This chapter advance the notion that there are six dominant varying types of algorithms, and the way they are used. The opportunities and risks associated with the use of these algorithms in decision-making depend on understanding the context in which an algorithm function.

The domain of Artificial Intelligence has advanced many techniques to automate the process of cognitive analysis and decision-making with distinctive attention paid to situations of high uncertainty. This chapter examines the basic decision-making concepts and algorithmic pathways related to topics such as logic, constraint modeling, and probabilistic modeling, as well as examining new research that utilizes these tools for predictive modeling and decision-making. An algorithm is a process or set of rules to be followed in calculations or other problem-solving operations, especially by a computer. Further, machine learning is a set of algorithms that enables the software to update and "learn" from previous outcomes without the need for programmer intervention. It is fed with structured data in order to complete an assignment without being programmed how to do so.

Artificial Intelligence-related technological areas such as machine learning, neural networks and deep learning algorithms have become part of our daily lives. This technology is implemented in mostly every field of endeavor such as translation services, improving medical diagnostics, personal and enterprise banking and creating computer models in climate science.

Artificial Intelligence denotes cognitive processes, specifically related to reasoning, i.e., Artificial Intelligence represents machine intelligence or a machine's ability to reproduce some of the cognitive functions of a person. It has the capability to learn and solve problems. Moreover, before making any decision choice, individuals also reason; therefore, it is natural to explore the links between Artificial Intelligence and decision-making.

When decision-makers and organizational executives have reliable data analyses, recommendations and follow-ups through Artificial Intelligence systems, they can make enhanced decision choices for their organization, community, stakeholders and employees. Artificial Intelligence not only improves productivity of the individual team members, but also improves the competitive standing of organizations.

Prior to the resurgence of Artificial Intelligence and its ensuing commercial application, managers had to rely on inconsistent and incomplete data. With Artificial Intelligence, they have databased models and simulations to turn to for decision-making purposes. The limitless outcome modeling is one of the breakthroughs in today's Artificial Intelligence systems. In other words, there is a gigantic opportunity to use Artificial Intelligence in all types of decision-making.

Advantages of Artificial Intelligence Tied to Decision-Making

The advantages of providing Artificial Intelligence a role to play in organizational decision-making are numerous:

- *Speedier decision-making*: The tempo of organizations has hastened and displayed no sign of decelerating in the global community. Further, the ability to speed up the decision-making process is imperative. For example, petroleum companies, retailers, travel sites, and other services routinely use dynamic pricing to optimize their margins.
- *Improve management of multiple inputs*: Machines are far better than humans at handling many distinctive factors at once when making multifaceted decisions, can process much more data at once, and use probability to imply or implement the best possible decision choice.

- *Reduction of decision fatigue*: Typically, people are compelled to make multiple decisions over a short period. Due to time pressures, the quality of those decisions may depreciate over time. Dissipated by all the decisions made during a shopping trip, consumers find it much difficult to defy the enticement of a sugar rush at the point of sale (perhaps this is why supermarkets place candy and snacks at the cash counters). Algorithms have no such weaknesses and can assist decision makers' escape from making inadequate decisions borne of exhaustion.

- *More innovative thinking and non-intuitive predictions*: Artificial Intelligence assists decision makers notice configurations that may not be readily ostensible to human analysis. For instance, a notable pharmacy ascertained through Artificial Intelligence that people who bought beer also tended to buy diapers at the same time (https://www.theregister.co.uk/2006/08/15/beer_diapers/). This type of unique discernment can have immediate influence on an organization.

There are several Artificial Intelligence applications which can heighten decision-making competences. Some of them are:

Automation efficiency and Artificial Intelligence

The automation efficiency advanced by Artificial Intelligence to modern day processes has gone beyond the assembly lines of the past. In several business functions, such as distribution, Artificial Intelligence has been able to hasten processes and provide decision-makers with reliable insight.

Moreover, distributive automation with the assistance of Artificial Intelligence has also been a key advantage of several retailers. Through Artificial Intelligence-supported monitoring and control, retailers can accurately predict and respond to product demand.

Customer Relationship Management (CRM)

Artificial Intelligence within CRM systems empowers its many automated functions, such as contact management, data recording and analyses, and lead ranking. Artificial Intelligence's buyer characteristic modeling can also provide an organization with a prediction of a customer's lifetime value. The sales and marketing teams can work more competently through these features.

Marketing decision-making with Artificial Intelligence

In marketing, the automation of market segmentation and campaign management has empowered more efficient decision-making and rapid action. Invaluable insight is obtained on customers, which can assist in obtaining improved interactions with them. Marketing automation is one of the main features of a good CRM application.

Further, there are various intricacies to each marketing decision. A decision maker has to know and understand the customers' wants, needs and desires, and associate services/products to them. Similarly, in the short- and long-run having a good comprehension of changing consumer behavior is critical to producing the best marketing decisions.

Artificial Intelligence modeling and simulation techniques formulate reliable insight into an organization's customer characteristics. These tools can be utilized to predict consumer behavior through an organizational network, an Artificial Intelligence system is able to support decisions through real-time and up-to-date data gathering, forecasting, and trend analysis.

Social computing

Social computing helps marketing professionals understand the social dynamics and behaviors of a target market. Through Artificial Intelligence, they can simulate, scrutinize and ultimately predict consumer behavior. These Artificial Intelligence applications can also be implemented to comprehend and data-mine online social media networks.

Opinion mining

Opinion mining is a type of data mining that searches the web for opinions and feelings. It is a way for marketers to know more about how their products are accepted by their target audience. Manual mining and analyses necessitate several hours. Artificial Intelligence has helped cut back this through reliable search and analysis functions.

This form of Artificial Intelligence is frequently utilized by search engines, which repeatedly rank an individuals' interests in specific web pages, websites and products. These bots engage different algorithms to get to a target's HITS and PageRank, among other online scoring systems. At this juncture, hyperlink-based Artificial Intelligence is exercised, wherein bots obtain clusters of linked pages and see these as a group sharing a common interest.

Online news and social media

Artificial Intelligence is breathing life into areas such as online news and social media monitoring by deepening its analysis of the billions of unstructured external data available beyond an organization's own four walls. All this unstructured data comes from millions of publicly accessible documents, social media feeds, digital pictures, videos, audio, and content on the web.

With the use of Artificial Intelligence, these unstructured data insights outside an organization can now be evaluated faster and more accurately to bring up-to-date the strategic direction of a brand. Moreover, Artificial Intelligence adds yet another level of analysis to media monitoring, permitting organizational leaders to turn their focus and time towards applying these insights in more creative and driven manners.

Failure to understand and adopt this modern mental framework towards media monitoring will risk falling behind their competitors or squandering time on menial tasks that proves to be an obstacle to more important organizational goals.

Recommendation system

Implemented in music content sites initially, recommendation system has been extended to different industries. A recommendation system is a subclass of information filtering system that search for "rating" or "preference" a decision maker would give to an item (Ricco and Shapira, 2011). Further, the Artificial Intelligence system learns a user's content preferences and drives content that fit those preferences. Further, the information learned by the Artificial Intelligence system can fashion better targeted content.

Machine learning algorithms in recommender systems are typically classified into two categories—content based and collaborative filtering methods; however, modern recommenders combine both approaches. Content based methods are grounded on similarity of item attributes and collaborative methods, which calculate similarity from interactions.

Expert system

Expert systems are a collection of rules programmed by individuals in form of if-then statements. Moreover, it is not part of Artificial Intelligence since they lack the capability to learn autonomously from external data. Nonetheless, Artificial Intelligence has tried to imitate the knowledge and reasoning methodologies of experts through expert systems, a type of problem-solving software. Expert systems, such as the marketing type MARKEX (Matsatsinis and Siskos, 1999), employ expert thinking processes to provided data. Output includes assessment and recommendations for a specific problem. This system behaves as a consultant for marketeers, extending visual support to enhance understanding and to vanquish a lack of expertise. The databases of the marketing system are a product of consumer surveys, as well as financial information of the organizations that is part of the decision-making process. The marketing system's model base encompasses statistical analysis, preference analysis, and consumer choice models. Finally, MARKEX includes partial knowledge bases to support decision makers in discrete stages of the product development process.

Yet another example is the online retail giant, Amazon. In 2012, Amazon acquired Kiva Systems, which created warehouse robots (https://pitchbook.com/news/articles/ma-flashback-amazon-announces-775m-kiva-systems-acquisition). In this enactment of robots, their mission was product monitoring and replenishment, and order fulfillment as well as lifting. The efficiency of these robots are custom-made to zoom through warehouses, fulfilling shipment requests at a very rapid pace.

Artificial Intelligence systems can deal with the enormous amount of data currently available. There is nothing humanly possible that can sift through this amount of data, in order to make it useable for business. Big data represents the most abundant, valuable and complex raw material in the world. And until now, we have not had the technology to mine it.

Throughput Modeling Algorithms

One of the principal problems of Artificial Intelligence is making the source code transparent when compared with other factors in algorithmic functioning. More to the point, machine learning algorithms and deep learning algorithms are typically built on just a few hundred lines of code. The algorithmic logic is generally learned from training data and is not often mirrored in its source code. That is, some of today's best-performing algorithms are more often than not the opaquest. Elevated transparency might entail deciphering a great deal of data, and then still only being able to fathom at what lessons the algorithm has learned from it.

Depending on the depth of a deep learning algorithmic neural network, there could be limited explanation and transparency. While an individual may be able to understand a machine with four or even 40 gears, levers and pulleys, most would struggle to explain something with 400 moving parts. This black box problem of Artificial Intelligence is not new, and its significance has grown with modern, more powerful machine learning solutions and more state-of-the-art models. A way to maneuver over deep learning's impenetrability is to forward a decision-making model to overview the playing field. That is, users typically will not trust black box models, even though they do not need or desire high levels of transparency. Instead, organizations should work to provide basic insights on the factors driving algorithmic decision choices.

Therefore, a Throughput Model is discussed in this chapter, which includes parallel processing of individuals' perceptions (P), information

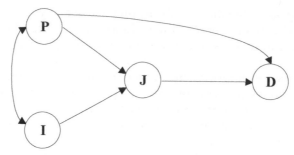

Figure 2.1. Decision makers' algorithmic processes. Where P = perception, I = information, J = judgment, and D = decision choice

sources (I), judgments (J), and decision choices (D) (see Figure 2.1). The model produces several salient premises regarding individuals' parallel processes implemented to allow relevant and reliable information to be accessed when required to make a decision/choice. There are a number of assumptions fundamental to the view that is proposed in this chapter. First, a number of different perceptual and informational parallel cues can retrieve any event relevant to the judgment process. Second, perceptual and judgmental processes have different parallel pathways leading to an action. Finally, the choice of retrieval cues is highly sensitive to the particular processing methods that are used by decision-makers.

There are several kinds of algorithms with differing functions and concomitant issues of bias and transparency. The Throughput Model provides six dominant algorithmic pathways which arrange the major concepts of perception, information, judgment and decision choice differently. That is, each pathway is in a specific algorithmic form, which sometimes means changing the data to match the algorithmic pathways' requirements. Data manipulation does not change the content of the data. What it does is change the influences from one step to another, presentation, and form of the data whereby an algorithm can help assist a person to see new patterns that were not apparent before in terms of ethics, trust and decision-making processes.

Based upon the Throughput Model, these algorithms are:

(1) P → D
(2) P → J → D
(3) I → J → D
(4) I → P → D
(5) P → I → J → D
(6) I → P → J → D

The Throughput Model (1984, 1991a,b, 1997, 2006, 2016) can be viewed as a depiction of human and artificial intelligence, which explains how a decision choice is made and which factors are involved. In Figure 2.1, the 'P' represents perception, the 'I' portrays information, the 'J' for judgment, and the 'D' represents decision choice. This figure does not have one exact way for showing how a decision is made because there are several different ways to come to a decision and everyone will interpret the information differently. Someone's perception can be influenced by their previous experiences, their background including their education. Information will always be the same information provided to every individual but based on their perception they can process the information differently. Some people might have the ability to make further assumptions based on the knowledge they already have. Therefore, all these factors will lead to everyone's judgment being distinct from one another. Based on their judgement, everyone's decision will be different, it does not necessarily mean it is right or wrong, but they have different reasons for their decisions.

The efficiency of algorithms is in part, contingent on data. As organizations proliferate their use of data, the transparency of the data on which the algorithmic decisions are being made is critical to ensure accountability. The Throughput Model's algorithmic pathways provide many opportunities for organizations to be transparent about the data. Along with this modeling perspective, another feature is to communicate the context in which the data is being implemented, the quality of the data, its representativeness, and possible limitations. Further, it is essential to document how data has been collected and analyzed. For example, metadata should include information on how data is processed and interpreted: communicating the assumptions behind aggregation, abstraction, knowledge and hypothesis makes it possible to reexamine data to discover things that were not initially noticed (UK Government Office for Science, 2016). Finally, documenting the kind of data and how it has been used to inform decision-making can assist to diminish any uncertainty on how decision choices are made, and to allow for decisions to be questioned.

Understanding that the Throughput Model is a representation of the human mind allows for a depiction of at least two-dimensional views of Artificial Intelligence. The first dimension connects Artificial Intelligence to the sciences of the artificial (Simon, 1969). This viewpoint can also depict the science of designing and building computer-based objects performing a variety of human chores (Pomerol, 1997). This particular perspective of Artificial Intelligence has relatively not many connections with decision-making to the boundary that an object cannot properly be said to make a decision choice. That is, when a task is programmed, the decision choice no longer exists since the actions are determined corresponding to each possible situation that may occur.

The second dimensional viewpoint relates to cognitive processing of individuals. Artificial Intelligence is sometimes viewed as the science of knowledge representation and reasoning (Newell and Simon, 1972; Rodgers, 1991). Hence, Artificial Intelligence is the science of the design and development of systems mimicking humans.

Can Artificial Intelligence Move Beyond Human Biases and Errors?

When coupled with a knowledgeable person, Artificial Intelligence can make its human counterpart faster and more efficient. In addition, Artificial Intelligence can facilitate individuals to reduce common or repetitive errors. Moreover, since world is full of information, our brains are only capable of processing a certain amount. If a person attempted to analyze every single aspect of every situation or decision, it would be very difficult to get anything done. In order to cope with the incredible amount of information that people encounter as well as to speed up the decision-making process, the brain relies on cognitive strategies or heuristics to simplify things. Hence, people do not have to spend endless amounts of time analyzing every detail.

Therefore, heuristics can play important roles in both problem-solving and decision-making. When we are trying to solve a problem or make a decision choice, we often turn to these mental shortcuts when we need a quick solution (Rodgers, 2006; Tversky and Kahneman, 1974). Heuristics are viewed as short-cuts in problem-solving or decision-making.

Heuristics can also play a role in *effort reduction*. Here, individuals use heuristics as a type of cognitive lethargy. Heuristics reduce the mental effort required to make decision choices. *Attribute substitution* is another way for calling on heuristics. This approach allows people to substitute simpler but related questions instead of more complex and challenging questions. Finally, heuristics can be activated in a *fast and frugal* manner, since they can be implemented expeditiously and can be correct.

At times, these heuristics can work quite well in problem solving or making decisions, mainly when a person is confronted with ill-structured information, changing environments, differential expertise levels, and time pressures. Nonetheless, there are times when the employment of these heuristics or strategies can spell disaster for a decision maker. Below are heuristics and biases that humans oftentimes find themselves employing when problem solving. Although, heuristics may speed up the problem-solving or decision-making process, they can introduce errors. Despite the fact that something has worked out well previously, does not necessitate that it will work again. Further, relying on an existing heuristic can make it difficult to see alternative solutions or come up with new ideas. Hence, heuristics can lead to inaccurate judgments regarding how common events happen and about how representative certain events may be.

People implement strategies to function generally in a systematic manner especially when confronted with what we believe to be similar events. Automatic strategies and information processing strategies are involved in most decision choices. Errors, biases, and context dependent strategies may result from mental mechanisms of which decision makers are largely unaware, and these may have a direct or indirect impact on decision choice.

Big data is exactly what it sounds like, a big amount of data. It represents data sets that are very big and complex that human minds and the technology in earlier years could not process or understand. The information could be misunderstood or not adequately processed because it was too much to comprehend. Now that technology has evolved and became something much bigger, researchers are trying to find a process to make information from big data useful and valuable. There is a number of ways that this can be done but they have not found the most practical way of doing this. The more technology advances the more information can be taken out of big data, the more information more relationships can be made.

Example: Decision Making Algorithms Representing Loan Officers' Processes

This section portrays the algorithmic causal chains of novices' and loan officers' processes through which financial statement information. Further, loan officers' perception of economic and management information is depicted as their knowledge representation and decision choices. Using both novices and experts provide a basis of comparison that can help us understand the intermediary processes of decision makers as part of Artificial Intelligence algorithms. The perceptual part of the algorithmic model captures the perceptual framing effects (i.e., P→J and P→D), which is not part of decision makers first stage analysis of company data. Instead, the perceptual stage represents cues, both external (e.g., economic events) and managerial (e.g., stewardship of a company), which may conflict with the actual financial statement information of a company. This procedure is partly motivated by the assumption that decision makers' processes are contingent on the decision problem as perceived by the decision maker. The objective of this investigation is to provide an Artificial Intelligence base from which to generate some useful suppositions about the factors that mediate novices' and loan officers' risk perceptions of information as well as their knowledge representation in judgments and decision choices. Therefore, in capturing an Artificial Intelligence algorithmic process, designers need to analyze the task to understand the domain-specific knowledge and strategies needed to perform the task. Researchers and practitioners can benefit from a decision-making model that can account for their information processing stages before a final decision is made. The Throughput Model provides a first step that may allow for an Artificial Intelligence computer simulation of the decision processes of users and preparers of conflicting economic, management, and financial information (see Figure 2.2).

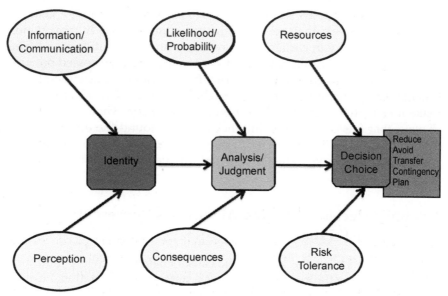

Figure 2.2. Three key steps in risk management decision choices. Source: Adopted from Rodgers (2006)

In the remaining sections, an Artificial Intelligence discussion of portraying the effects of information and perception on judgment and decision choice is presented. Also, the decision makers' processing Throughput Model is presented along with exploration and examination premises for decision makers.

The decision makers' processing model

Prior studies (e.g., Rodgers, 1984, 1991b, 1999; Rodgers and Housel, 1987, 2004) concluded that the following significant relationships exist in decision making tasks: in the first algorithmic stage, perception and information influencing judgment (i.e., P→J; I→J); while in the second algorithmic stage, perception affects decision choice and judgment affects decision choice (i.e., P→D; J→D).

The Throughput Model demonstrates that decision makers use whatever combination of different information algorithmic processing pathways and strategies they perceive as appropriate for the decision at hand (see Rodgers, 1991b). Moreover, some decision makers may have strategies that resemble traditional decision theoretical models, but most do not. Decision-making models ought to focus on continuous predictions occurring in dynamic and complex task environments. Adoption of such a framework can closely approximate real-world decision making. The approach used in this section captures a simulated continuous and dynamic laboratory task environment which could lead to new insights about decision-making in an individual or organizational context (see Figure 2.1).

An algorithmic two-stage loan making task is symbolic pertaining to what is used in strategic organizational decision-making processes. That is, loan officers' decision processes are influenced by very similar educational backgrounds; specifically, in the kinds of training they receive to qualify as loan officers. The first stage of their training covers certain conditions that are necessary before the actual analysis of the management information begins (second stage). These conditions represent how they perceive economic, management and financial information of an organization. Also, these conditions are necessarily based upon the credit policy, philosophy, and procedures of a lending institution. Categories of 'granting a loan' and "not making a loan" could be based upon these representations (Rodgers and Housel, 1987). Therefore, decision makers' first stage of processing includes their representation of economic and management information. The first stage of processing also includes considering information which may represent financial statement information. These three informational sources (i.e., economic, management and financial) discussed in the next section are part of decision makers' preliminary credit decisions and may impact on their information analysis (i.e., second stage of processing), thereby influencing their loan granting decision (Rodgers, 1991b).

Decision makers' economic risk perception includes characteristics of the industry (e.g., susceptibility to economic fluctuations), economic climate (e.g., forecasts of the gross national product), company forecasts (e.g., product innovation), and government regulations (e.g., potential impact on industry and company). Their management risk perception includes management capability (e.g., well defined objectives and goals), adequate controls (e.g., timely, consistent monitoring and measurement of progress towards identified objectives), and sound organization and adequate depth (e.g., plan to develop future management). Financial information may be measured by liquidity (e.g., current ratio), income (e.g., net margin ratio), and risk (e.g., debt/net worth ratio).

Knowledge structures in algorithms

A central theme in this section is the postulation of interacting knowledge structures, which is designated as schemata. Schemata are data structures for representing the broad concepts stored in memory. They exist for generalized concepts underlying information, situations requiring analysis, decision choices, and sequences of choices. Representations like schemata are useful structures for encoding knowledge in decision-making activities. Instinctively, these activities appear to require mechanisms in which each aspect of the information in a particular situation can act on other perceptual processing, simultaneously influencing judgmental processing. To convey these intuitions, machine learning, neural networks, deep learning techniques assume that information processing takes place through the interactions of a large number of simple processing elements called units, each sending activations to other units. In some situations, the units stand for possible premises about such things as the loan officers' perceived ideas about a possible loan strategy for a customer, or their predictive judgments about a customer's ability to finance a particular project (Rodgers, 1991b). In such situations, the activations stand approximately for the strengths associated with the different possible hypotheses.

The Throughput Modelling approach offers the hope of computationally sufficient and psychologically accurate mechanistic accounts of the phenomena of decision makers' cognition which have eluded successful explication in conventional computational formalisms; and they have radically transformed the manner we think about the time-course of processing, the nature of representation, and the mechanisms of learning. In the next section, we propose a Throughput Model which takes the serial processing model one step further by capturing decision makers' perceptual and judgmental parallel processing before they make a final decision.

The Throughput Model algorithmic approach

The primary Artificial Intelligence goal is to construct a cognitive architecture inspired by cognitive psychology. That is, the human brain is not just a homogenous set of neurons. Further, knowledge based on general facts (declarative or semantic memory), as well as knowledge about previous experiences (or personal facts) is called episodic memory (Rodgers and Housel, 1992, 2004). There is a real structure in terms of different components, some of which are associated with knowledge about how to do things in the world. This process is referred to as procedural memory (Rodgers and Housel, 1992, 2004).

In this section, a Throughput Model is proposed to predict and explain individuals' decision choices. If knowledge representation is an essential determinant of how a person represents a problem and utilizes information, it is pertinent to know what governs the functionality of perception in the model. Information is coherent with perception in the model since it is subjectively processed by decision makers (Rodgers, 1984; Rodgers and Housel, 1987; Rodgers, 1991). Information is characterized by different information sources. These concepts are essential since decision makers generally rely upon certain types of information when they are considering an action (Rodgers and Johnson, 1988). Information affects judgment since it is stored in memory and affects summary inferences of individuals. Information and the decision maker's prior expectations or beliefs about the information are relevant to perceiving the degree of co-variation between them. Moreover, information sources jointly determine co-variation with perception. Knowing that certain prior expectations tend to be associated with informational sources allows one to prepare for the future and behave appropriately. One way to view predictive ability and the knowledge representation upon which it is based is to consider that it results in part from assessments of the co-variation of information derived from experience. The information sources in the Throughput Model also affects judgment. For example, information processing takes place through the interactions of many informational units, each sending activation signals to the judgment unit. The units stand for possible assumptions about what information units have a significant effect (i.e., activation) on the judgment concept in analyzing a potential analysis of a situation. The activations stand roughly for the strengths associated with the different possible information sources.

After the initial perceptual processing is completed, a decision maker derives summarizing inferences or judgments in order to make his/her final decision. Kahneman and Tversky (1979) have used prospect theory to describe this categorization-inference process as the testing and confirming of a decision maker's selected information (i.e., framing of decisions). Finally, perceptual processing of a decision maker can bypass the summary inferences and directly affect his/her final decisions. That is, users of information may rely upon prior similar information to influence their decisions without going through an in-depth analysis of the presented information (Rodgers, 1999).

A number of experiments have been aimed at an explanation of judgmental bias (e.g., Rodgers, 1991b, 1999). These results indicate that a major cause of judgmental bias is the decision maker's mis-aggregation of the data. That is, decision makers perceive each datum accurately but are unable to combine its diagnostic meaning well with the diagnostic meaning of other data when revising their opinion. Some researchers have indicated that decision makers' pattern recognition method of processing information can have a biasing effect on their perceptions of information (e.g., Rodgers and Johnson, 1988; Rodgers and Housel, 2004). For example, there is some evidence that suggests that certain individuals perceive large configurations more accurately than they perceive its components (top-down processing), whereas other individuals do just the opposite (bottom-up processing) (Rodgers and Housel, 1987; Rodgers, 1992). Bottom-up processing, in which perceptual units combine to form larger units, is sometimes referred to as data-driven processing. In this kind of processing presented information controls or directs an individual's processing. For example, net income from a set of accounting information would greatly influence how an individual process and analyzes larger components of the information set.

Top-down information such as context and general knowledge, is sometimes referred to as conceptually-driven processing. In such processing, the perceiver imposes his knowledge and conceptual structures on the accounting information set to help decide what type of organization is being represented. Prior research (e.g., Rodgers and Housel, 1987; Rodgers, 1992) has shown that conservatism may arise more so with this type, since the data may be mis-aggregated when the diagnostic meaning of the data is combined together. Conservatism in our paper is defined as risk-averse behavior for loan-making purposes. Dividing decision makers into data-driven and conceptually-driven types can provide an anchor as a first approximation of their decision choice (Rodgers and Housel, 1987). This anchor or predispositional bias, is then adjusted to accommodate the implications of additional information to be processed by decision makers. Typically, the adjustment is imprecise and insufficient. For example, Tversky and Kahneman (1973) showed how anchoring and adjustment could cause the overly narrow confidence intervals found by many investigators and the tendency to misjudge the probability of conjunctive and disjunctive events.

Machine learning, neural network, and deep learning questions concerning the algorithmic pathways in the Throughput Model are discussed in the next two sections.

First stage: perception → judgment and information → judgment pathways

It is a well-known fact that we often misperceive unfamiliar objects as more familiar ones and that we can get by with lower-quality information in perceiving familiar items than we need for perceiving unfamiliar items (Rodgers, 2006). Not only does familiarity help us determine what the higher-level structures are when the lower-level information is ambiguous, it also allows us to fill in missing lower-level information within familiar higher-order patterns. This may have some implications that data-driven and conceptually driven processing are two different higher-level structures implemented when the lower-level information is ambiguous.

That is, algorithm selection is an important phase in understanding configurations in data. Further, designing a data-driven or conceptually driven model can improve predictability. Without tightening the model algorithmic design can influence the prediction models. Models are based upon the hyperparameters provided, which can lead to a myriad number of possible outcomes. Hyperparameters are lines of code reflecting the algorithm's setting. These lines of code are comparable to the controls on the dashboard of an airplane or the knobs utilized to tune radio frequency.

The perception and information schemata exemplify a process of information reduction whereby several variables are reduced into a simpler and more organized form. This process is similar to a neural network or Bayesian statistical analysis. These organizational processes result in, for example, loan officers' perceptions being structured into units corresponding to financial and non-financial statement information and properties of that information. It is these larger units that may be stored and later assembled into judgment knowledge. That is, in the Throughput Model, knowledge is derived partially from perceptual representations and information sources. Our machine learning question addresses whether judgment is significantly affected by the parallel processing of perception and information (see Figure 2.1). The activations of different parallel pathways may affect the way decision makers combine and utilize information in the second stage of processing. This first stage of processing undoubtedly has important indirect influences on decision makers' actions based upon their data-driven or a conceptually-driven orientation. To this end, decision trees, forest trees, regression analysis, and clustering analysis can be used to propel a data-driven or a conceptually-driven orientation.

Second stage: Perception → decision choice and judgment → decision choice pathways

Researchers (see Rodgers, 1991b) have argued that 'propositional' representation is suitable for representing knowledge. Propositions are abstract language-like representations that assert facts about decision-making tasks. This sort of representation is required to store decision-making information since what bankers know about granting loans to an organization is predicated upon a set of 'causal assertions' that are necessarily either true or false (Rodgers, 1984, 1991, 2017; Rodgers and Johnson, 1988). The concept of 'causation' appears to correspond to reasoning schemas (Rodgers and Housel, 1987). There are very likely several kinds of causal schemas, adjusting on whether single or multiple causes are believed to create the effect. That is, multiple causes of decision choice are believed to be from perception and judgment. Since different processing schemas can be inferred as a result of decision makers having a data-driven orientation or a conceptually driven orientation, a fundamental question arises: Do both types' perceptual and judgmental processes have a direct influential pathway to decision choice? It is our view that the answer to this question has essential inferences for understanding the nature of machine learning training and may further explain the knowledge representation cues retrieved by Artificial Intelligence algorithms engaged in hypothesis confirming or hypothesis-disconfirming search strategies (Rodgers, 1991).

Many studies of organizational performance (e.g., Foss and Rodgers, 2011) and studies of fitting linear models to decisions made by loan officers (e.g., Rodgers and Johnson, 1988) implied a definite potential of ratios as predictors of successful and unsuccessful organizations. The Throughput Modelling process attempts to address the aforementioned in three ways. First, the research design incorporates loan officers' use of their perceptual and judgmental parallel processes as schemas to select financial statement information prior to making decision choices. Loan officers' perceptions indicate their classifying and categorizing of information (based upon previous experience with loan applicants) before they perform a detailed analysis of the present loan prospect.

Second, data-driven and conceptually-driven orientations, which are two types of perceptual schemas, can be utilized in a machine learning setup using decision trees or neural networks if: (1) they select distinct information in their analysis, and (2) one type outperforms the other. The overall objective is to determine if loan officers would implement certain types of schemas (i.e., data-driven or conceptually-driven) and whether the use of different information cause them to be more conservative in their decision choices. Finally, the Throughput Model adds to Artificial Intelligence employed machine learning, neural networks, or deep learning by measuring the effects of information schemas regarding financial and non-financial information influences on loan officers' perceptions and judgments before they make a decision choice.

To summarize, Artificial Intelligence questions relating to modeling behavior can be addressed in 'two stages' in the Throughput Model. The first stage represents parallel processing effects of perception and information on judgment. The second parallel processing stage depicts effects of perception and judgment on decision choice.

Conclusion

In the end, all technology revolutions are driven not only by discovery, but also by organizational and societal necessities. These new possibilities are pursued not because we can, but instead because we must.

This chapter illustrated how decision makers' algorithmic processes can influence their analysis, which in turn can impact on their decision choices. This chapter also implies that users of information generally selected information in order to come to a decision point. But before users of information come to a decision, there are certain algorithmic pathways that may influence their final choice. That is, knowledge representation of information represents users' perceptions; whereas, in the second stage of processing inferences and analysis of information represents their judgments. It is important to understand decision makers' knowledge representation in the first and second stage of processing in order to explain and predict what types of information are of importance before a final decision is made. Also, if decision makers utilizing information are aware of their biases, they may be able to correct and modify their future actions.

Moreover, in this chapter, the Throughput Model depiction of the cognitive view of knowledge representation was used as a basic ingredient in understanding how loan officers may implement parallel processing when analyzing information. This perspective was used since it offers an approach of how decision makers process information in the first stage (i.e., perception and information) and the second stage (i.e., perception and judgment) towards their decision choice. The Throughput Model permits decision-making behavior to be analyzed and decomposed into a set of labelled constructs (financial and non-financial information, perception, judgment, and decision choice). In this format, decision-making behavior involves activating financial and non-financial information through a set of two stage operations into a decision choice.

Previously, pattern recognition was depicted as resulting from bottom-up information (data-driven) from data sources and top-down information (conceptually-driven) from context and expectation. One way of portraying knowledge representation is to say that we combine information from these multiple sources to deduce what we select to choose. Nonetheless, particular types of decision makers are noted as having a perceptual bias of being either data-driven or conceptually-driven toward information. Conceptually-driven types are more conservative about their loan decisions than the data-driven types. Since these types perceive larger configurations more accurately than they perceive its components, they may have reaggregated the data. That is, this type may be able to perceive each datum accurately and are well aware of its individual meaning; however, it did not combine as well its diagnostic meaning with the diagnostics of other data. This type of processing not only influences decision makers' perception, but also their judgments. That is, conceptually-driven types' selection of information and their inferences are to some extent different from the data-driven types.

Understanding decision makers' perceptual biases is important in how machine learning algorithms categorize and make inferences from information items. That is, data-driven types may have a more complicated form of information processing. Their use of more parallel cues provided them with a broader knowledge representation of a loan prospect (Rodgers, 1991). Algorithms employed in Artificial Intelligence modeling should be concerned if conceptually-driven types' biases, as well as their conservatism based upon mis-aggregated data, can be controlled for rational decision-making purposes. Further, machine learning techniques should also acknowledge different types of direct and indirect financial and non-financial information effects on decision choices as a reflection of the Throughput Modelling two-stage processes.

Governments, businesses and public bodies will need to deliberate the use of algorithms in decision-making, consulting widely, and ensuring that mechanisms are in place to detect and address any mistakes or unintended consequences of decision choices made. Artificial Intelligence is becoming more pervasive in most aspects of decision-making in the foreseeable future. When decision-makers and organizational executives have reliable and relevant data analyses, recommendations and follow-ups through Artificial Intelligence systems, can make better choices for their organizations and employees.

Nonetheless, for an algorithmic decision-making process, there are complex challenges to reach an informed decision. For example, sometimes it is difficult to know whether decision-making algorithms will be able to make effective and efficient decision choices with the current computing and data analytics infrastructure and processing capability.

In addition, while the regulatory landscape is developing, governments, businesses and public entities should lead by example by applying standards to its own use of algorithms, to ensure accountability and help build public trust in use of algorithms. Further, it is important to contemplate the protection of personal data, auditability, and liability for harm caused by the use of algorithms. The Throughput Model's algorithms provide an elucidation for the aforementioned issues.

Artificial Intelligence enhanced decision-making model will eventually be the driver of the future global economy. Systems will enable autonomous decision-making that are complex in dynamic environments, such as financial services, logistics, ride sharing and autonomous vehicles, port management and other autonomous systems. The Throughput Model presented in this chapter is based on theoretical underpinnings that depict perceptual framing and use of information in a decision-making dynamic environment.

Providing understandable and explainable models is a necessary condition for sophisticated machine learning and deep learning models in an area of research and practice. The Throughput Model provides for at least five approaches to assisting users and developers of sophisticated machine learning and deep learning models. These are:

1. *Straightforward conceptual modeling*: This may relax exactness for explainable modeling.

2. *Combine the Throughput Model with more sophisticated models*: The more sophisticated models offer the recommendation, the Throughput Model provides the rationales. This can work quite well; nonetheless, there may be gaps when the models deviate.

3. *Use of intermediate stages in the model*: The Throughput Model algorithmic pathways provide for intermediary stages (e.g., judgment) that are excited by certain patterns. These can be visualized as features tantamount to compensatory versus non-compensatory analysis in order to provide a rationale for classification.

4. *Use perceptual framing mechanisms*: Some of the most sophisticated models have a mechanism to direct "perceptual framing" towards the parts of the input that matter the most (i.e., setting higher weights). These can be envisioned to highlight the parts of an image or a text that contribute the most to a distinct recommendation.

5. *Modify inputs*: If striking out a few expressions or eliminating a few parts of an image significantly changes, overall results may indicate that these inputs display a significant role in the classification. They can be explored by running the model on variants of the input, with results that are underlined to the user.

Ultimately, individual/organizational decision-making can only be explained to some point. The same remains for sophisticated algorithms. Nevertheless, it is the software providers' responsibility to accelerate research on technical transparency to further construct transparency in the Artificial Intelligence software.

Further, the Throughput Model contrasts the differences decision makers that use different pathways or strategies in problem solving. Moreover, the Throughput Model highlights a two-stage process that may enable us to train data (e.g., supervisory or non-supervisory) in parts of the model within the context of the entire model. This is important since testing a part of a model separate from the entire model may result in different or bias Artificial Intelligence findings. That is, changes in a particular part of a model may affect other interrelated parts of the model (e.g., employment of decision trees). If we can understand how changes occur in the different parts of an archetype, then we may be able to design better Artificial Intelligence systems. The Throughput Model presented in this chapter addresses these challenges and, hopefully, future applications of the model will lead to even more fruitful applications of Artificial Intelligence decision-making concepts.

The basic premise of this chapter is that one way to gain an understanding of something is to build or model a likeness of it. In this sense, the financial and non-financial information people use is a model of their cognitive processes. Apparently, the ways in which decision makers make judgments and decisions are strongly influenced by their environment. Studying how decision makers happen to think in only a particular part of an environmental situation or at a particular time fails to cover the full range of possibilities of different algorithmic pathways to judgments and decisions.

Decision-making using financial and non-financial information involves a search process. The critical aspect of this search process is the decision maker's mental representation of the goal, whether it be a possibility or a piece of evidence. The goal provides the most essential direction. For example, if the goal of a loan officer is to make an unsecured loan to a particular organization, then she/he tries to find different possibilities than if the goal is to make a secured loan. Loan officers in their search for supporting evidence either for or against an unsecured loan, might direct their search for new ones. Artificial Intelligence supported machine learning, neural networks, and deep learning apparatuses should address more "foolproof" strategies for decision makers using such systems. Then we could concentrate on addressing anomalies.

A strong decision-making theory can assist in deployment of various technology-based apparatuses that underlies Artificial Intelligence-enabled tools combined with other disciplines (e.g., robotics, human-computer interaction, computer vision and the humanities and social sciences). This approach should take into account how these intelligent systems interact and collaborate with humans and consider their validation and verification, especially in application areas where the dependability, ethics, trustworthiness, safety or security of implementations is a concern. Artificial Intelligence researchers will play a key role in furthering future intelligent technologies and data enabling decision-making cross a variety of platforms and are well-placed to contribute to the other cross technologies. In order to maximize the impact of these contributions, a decision-making theory should inspire effective communication with researchers and practitioners in other contributing areas such as natural language processing, visualization, healthcare, policy setting as well as institution building.

References

Kahneman, D. and Tversky, A. 1979. Prospect theory: An analysis of decision under risk. *Econometrica*, 47: 263–291.

Foss, K. and Rodgers, W. 2011. Enhancing information usefulness by line managers' involvement in cross-unit activities. Organization Studies, 32: 683–703.

Matsatsinis, N.F. and Siskos, Y. 1999. MARKEX: An intelligent decision support system for product development decisions. European Journal of Operational Research, 113(2): 336–354.

Pomerol, J.C. 1997. Artificial Intelligence and human decision making. European Journal of Operational Research, 99: 3–25.

Ricco, F. and Shapira, B. 2011. Introduction to recommender systems handbook. Recommender Systems Handbook, pp. 1–35. NY: Springer.

Rodgers, W. 1984. Usefulness of decision makers' cognitive processes in a covariance structural model using financial statement information. University of Southern California, Dissertation.

Rodgers, W. and Housel, T. 1987. The effects of information and cognitive processes on decision-making. Accounting and Business Research, 69: 67–74.

Rodgers, W. and Johnson, L. 1988. Integrating credit models using accounting information with loan officers' decision processes. Accounting and Finance, 28: 1–22.

Rodgers, W. 1991a. Evaluating accounting information with causal models: Classification of methods and implications for accounting research. Journal of Accounting Literature, 10: 151–180.

Rodgers, W. 1991b. How do loan officers make their decisions about credit risks? A study of parallel distributed processing. Journal of Economic Psychology, 12: 243–265.

Rodgers, W. 1992. The effects of accounting information on individuals' perceptual processes. Journal of Accounting, Auditing, and Finance, 7: 67–95.

Rodgers, W. and Housel, T. 1992. The role of componential learning in accounting education. Accounting and Finance, 32: 73–86.

Rodgers, W. 1997. Throughput Modeling: Financial Information used by Decision Makers, Bingley: Emerald Group Publishing.

Rodgers, W. 1999. The influences of conflicting information on novices' and loan officers' actions. Journal of Economic Psychology, 20: 123–145.

Rodgers, W. and Housel, T. 2004. The effects of environmental risk information on auditors' decisions about prospective financial statements. European Accounting Review, 13: 523–540.

Rodgers, W. 2006. Process Thinking. Six Pathways to Successful Decision Making. NY: iUniverse, Inc.

Rodgers, W. 2012. Biometric and Auditing Issues Addressed in a Throughput Model. Charlotte, NC: Information Age Publishing Inc.

Rodgers, W. 2017. Knowledge Creation: Going Beyond Published Financial Information. Hauppauge, NY: Nova Publication.

Simon, H.A. 1969. The Sciences of the Artificial. Cambridge, Mass: MIT Press.

Tversky, A. and Kahneman, D. 1973. Availability: A heuristic for judging frequency and probability. Cognitive Psychology, 5(2): 207–232.

Tversky, A. and Kahneman, D. 1974. Judgment under uncertainty: Heuristics and biases. Science, 185: 1124–1131.

UK Government Office for Science. 2016. Artificial intelligence: opportunities and implications for the future of decision making. www.gov.uk/government/uploads/system/uploads/attachment_data/file/566075/gs-16-19-artificialintelligence-ai-report.pdf.

3

Artificial Intelligence

Six Cognitive Driven Algorithms

"Education is the most powerful weapon, which you can use to change the world."

—Nelson Mandela

"Success in creating effective Artificial Intelligence, could be the biggest event in the history of our civilization. Or the worst."

—Stephen Hawking, Physicist

Today, quite a few decision choices can be made by computer algorithms ranging from interpreting medical images to recommending books or movies. These Artificial Intelligence advanced analytic capabilities can have access to large amounts of data. The growing prevalence of these algorithms has led to widespread concerns about their impact on those who are affected by decisions they make. To the champions of this Artificial Intelligence movement, these systems promise to increase accuracy and reduce human bias in important decisions. Nevertheless, others are concerned that many of these Artificial Intelligence systems amount to "weapons of math destruction" that simply reinforce existing biases and disparities under the guise of algorithmic neutrality.

Therefore, one of the fundamental questions when it comes to Artificial Intelligence is: Are we going to implement Artificial Intelligence to make decision choices? If that is the case, are we going to use it to support individuals' decision-making? Are we going to implement Artificial Intelligence in order to make decision choices autonomously? If so, what are the weaknesses and pitfalls of this approach? What are the positives to using Artificial Intelligence? And how do we manage this process?

We know Artificial Intelligence has a great deal of possibilities; however, what will be the learning curve for society. What decision-making pathways do we implore, whereby we can gain control over the narrative of how Artificial Intelligence tools influences the decision choices that are made for us or about us?

When solving a problem, selecting the appropriate pathway is often central to reaching the suitable solution. One of these problem-solving approaches is known as an algorithm. An algorithm is a defined set of step-by-step procedures that provides the appropriate response to a particular problem.

By appropriately following the instructions, a person is guaranteed to arrive at a correct answer. While often thought of purely as a mathematical term, the same type of process can be charted to warrant finding the suitable answer when solving a problem or making a decision choice.

Let's say I start with two slices of bread, spreading creamy garlic basil on one slice and mayonnaise on the other slice. I put a slice of vegan cheese on the bread with the mayonnaise, some veggie bacon on top of that, some lettuce, two slices of tomato and then top it with that slice with the creamy garlic basil on it.

Definitely I eat it right away. But if I leave it on the table for a while, that top slice of bread may become moist from being drenched in the tomato. Therefore, it may be a problem I did not recognize earlier, and I might have been making sandwiches for years before becoming aware of it. However, once I do, I can start thinking of ways to alter my algorithm in order to create an improved sandwich.

For example, I could eliminate the tomato. Yet, I do not want to miss the taste of tomato. Therefore, as an alternative, I can put the tomato on the sandwich after the bread and the lettuce. This permits the lettuce to form a shielding barrier between the tomato and the bread.

An algorithm is often expressed in terms of an equation, decision tree, graphs or pathways. For example, in terms of depicting a graph, a square could represent each step. Arrows then branch off from each step to point to possible directions that you may take to solve the problem. In some cases, you must follow a particular set of steps to solve the problem. In other instances, you might be able to follow different pathways (e.g., Throughput Model) that will all lead to the same solution.

For Artificial Intelligence technology to solve a problem, it has to be given a process or set of rules to follow, called algorithms. Since algorithms are fundamental to Artificial Intelligence technology, this chapter highlights the Throughput Model's six dominant algorithms. These six different sets of algorithms that portrays artificial intelligence consist of four concepts: perception (P), information (I), judgment (J), and decision choice (D). The algorithms have at least two concepts and a maximum of four concepts. Below are the six different algorithms, represented as the Throughput Model as follows:

(1) $P \rightarrow D$
(2) $P \rightarrow J \rightarrow D$
(3) $I \rightarrow J \rightarrow D$
(4) $I \rightarrow P \rightarrow D$
(5) $P \rightarrow I \rightarrow J \rightarrow D$
(6) $I \rightarrow P \rightarrow J \rightarrow D$

Machine learning emphasizes on the use of algorithms through which machines can learn by using data, observations and experiences. Powerful machine learning can assist in developing the six dominant algorithms in order to create various archetypes that can accurately predict future events and can provide notices to people in advance. It can also help machines to make intelligent decisions based on their past experiences. There is still a great deal of code to be written by researchers; hence, years of research is still required to reach a level where machines can learn by themselves and come up with solutions without human help based upon the six dominant algorithms.

Next is a discussion pertaining to six dominant algorithmic decision-making pathways which can be implemented for decision problems represented as multi-stage pathways. These algorithms construct scenarios incrementally, starting from problem framing that may or may not make use of the available information. The incremental process constructs a series of steps that can include information available to the decision maker depending on the algorithm used for the problem solving. While the algorithmic pathways converge to a decision choice, the Throughput Model approach is designed for situations in which determining the sources and strength of "perception" or "information" is feasible. This chapter is ordered in the following manner: (1) decision-making heuristics, (2) the six dominant algorithmic pathways, and (3) implementing a particular algorithm via type of information.

Heuristics Used in Problem-Solving and Decision-Making

Similar to the human brain, Artificial Intelligence is subject to cognitive bias. Human cognitive biases are heuristics, mental shortcuts that skew decision-making and reasoning, resulting in reasoning errors.

Some factors that can limit the ability to make sound decision choices entails noisy, missing or incomplete information, urgent deadlines, and limited physical or emotional resources. When making a decision choice, individuals configure opinions and select actions through cognitive processes that are swayed by biases, reasons, emotions, and memories.

The basic act of deciding substantiates the notion that we have our own preferences and determinations. People evaluate the benefits and costs of their decision choices, and then deal with the consequences (Rodgers, 2019).

Further, one of the most common basic elements in cognitive biases is inclination. Susceptibility in Artificial Intelligence is motivated through the allocation of weight on the parameters and nodes of a neural network, a computer system modeled on the human brain. The weight may inadvertently bias the machine learning algorithm from setting up through data input, through supervised training, and by involvement through manual adjustments. The absence or inclusion of indicators and the inherent cognitive biases of the human computer programmer can trigger machine learning bias.

Examples of cognitive biases include stereotyping, the bandwagon (i.e., herd mentality) effect, confirmation bias, priming, selective perception, the gambler's fallacy, and the observational selection bias (see Rodgers, 2006). Other heuristics and biases that individuals may employ include belief-bias, confirmation bias, as well as the anchoring/adjustment, representative, and availability (Table 3.1).

The *belief-bias* effect resonates due to our perceptual frame locked in a prior belief. That is, the belief in an idea, concept, or thing can overweigh other relevant and reliable information.

The *affect bias* comprises making decision choices that are strongly influenced by the emotions that a person is experiencing at that particular moment. For example, people may be more likely to see decision choices as having higher benefits and lower risks when they are in a positive mood. On the other hand, negative emotions may lead individuals to focus on the potential downsides of a decision choice rather than the possible benefits.

Table 3.1. Heuristics and biases (from Rodgers, 2006)

Hindsight bias	Ignore present information available at the time.
Belief bias	Resistance to new information (dogma).
Affect bias	Positive or negative emotions drive whether to accept a decision.
Confirmation bias	Selective framing of a problem whereby one tends to examine what confirms your beliefs.
Anchoring/ adjustment	Judgments are frequently influenced by a perceptual starting point.
Representative	We frequently perceive the likelihood of an event based on the similarity to the population from which it is drawn (i.e., our view of most typical situations).
Availability	Refers to estimates of frequency or probability that are made on the basis of how easily examples come to mind.

Confirmation bias refers to selective framing of a problem whereby one tends to examine what confirms your beliefs. In addition, one ignores or undervalues the relevance of what contradicts your beliefs. For example, if you believe that during a sunny day there is an increase in highway accidents, you will take notice of highway accidents on a sunny day and be inattentive to the accidents on a rainy day. A tendency to do this over time unjustifiably strengthens your belief in the relationship between sunny days and accidents.

The anchoring and adjustment strategy

This particular strategy is associated with judgments that are often motivated by a perceptual starting point and adjustments to that starting point. This strategy is typically related to facts and figures that can guide judgments. Notice the difference in your perception if I indicate to you that you can save $100.00 on the cost of a cellular telephone by buying it at a specific department store versus telling you that you can save $100.00 on the cost of a hi-tech home sound system by going to a certain merchandising store.

Your perception is anchored to the cost of these items. A $100.00 saving on a cellular telephone that costs $500.00 is perceived as significant compared to $100.00 savings on a home sound system that may cost about of $6,000. Nonetheless, in terms of your personal finances, $100.00 is $100.00, exactly the same amount no matter how it happens. If you desire it in one circumstance, you should want it likewise in the other.

Representative strategy

This strategy suggests that we habitually judge the probability of an event based on the similarity to the population from whereby it is drawn. Similarity in these circumstances is often defined in terms of prototype— our viewpoint of the most typical state of affairs.

Consider the set of results (Heads and Tails) from a coin toss experiment (Rodgers, 2006). Quite a few people perceive that the sequence H HH T TT is very normal to be a random event; it does not coincide with our view that for an event to be random, it must have a certain irregularity regarding it. Statistically, however, the sequence H HH T TT has the same probability as, say, HTTHTH.

Sample size and representativeness

Individuals often try to account for probability and statistics in their thinking and decision making. Regrettably, the probabilities and statistical conclusions we utilize are over and over again established on large sample sizes, regardless of the fact that our decision making comprises small sample sizes.

For example, you have tossed a coin ten times and it has come up heads all ten times. On the eleventh toss, is it more probable to come up a head or a tail? Most people have confidence in that it's more prone to come up a tail. Someone might say, "After all, it's come up heads ten times in a row. Unquestionably, it will come up tails next—it's the law of averages." As a matter of discourse, the probabilities of a head or a tail on the tenth toss are equal (0.5). The inappropriate conclusion that a tail is more possible is based on what is known as the *small-sample fallacy*, which presupposes that small samples will be representative of the population from which they are selected.

The small-sample fallacy could be, in part, one of the rationalizations for the compelling stereotypes we form of other ethnic and racial groups. Diversity and multiculturalism are operative ways of eradicating stereotyping centered on the small-sample fallacy; in that they facilitate us to become familiarized with large numbers of individuals from other groups.

Base rate and representativeness

Base rate proposes how often a feature (or event) occurs in the population. That is, the possibility of drawing such an item is equal to the number of such items divided by the total number of factors in the population. So convincing is representativeness and its accompanying stereotyping that we habitually ignore base-rate information and create our decision choices on representativeness. This is recognized as the *base-rate fallacy*. For example, the following personality summary of an imaginary person named Steve, which came out in a study by Kahneman and Tversky (1973, p. 241):

> "Steve is very shy and withdrawn, invariably helpful, but with not much interest in individuals or in the world of reality. A meek and tidy soul, he has a need for order and structure, and a passion for detail."

Subjects were invited to read these two sentences and then to judge Steve's occupation from a list of possibilities that encompassed farmer, salesperson, airline pilot, librarian, and physician. Knowledge of base rates would imply that there are a larger number of salespeople in the global community than there are librarians, leading to the conclusion that Steve is, in fact, most likely a salesperson. Nonetheless, since the description parallels to our stereotype of the librarian, most people think that he is a librarian. As demonstrated in studies by Amos Tversky and

Daniel Kahneman (1974), such errors happen even when the base rate is given.

The conjunction fallacy and representativeness

If two events transpire together or separately, the conjunction, where they intersect or (have something in common), cannot be more possible than the chance of either of the two individual events. Nevertheless, people forget this and ascribe a greater chance to combination events. That is, we incorrectly relate quantity of events with quantity of probability.

Given a question, which has the higher probability in a single draw from a deck of cards, acquiring a diamond or obtaining a diamond that is also a picture card? Nearly everyone has no obstacle with this problem. They appropriately reason that of the 52 cards in the deck, 13 are diamonds but only 4 are both diamonds and picture cards. Since 13/52 is greater than 4/52, clearly the probability of obtaining a diamond is higher than that of attaining a diamond that is also a picture card. Alternatively, consider the following demonstration constructed on an experiment by Tversky and Kahneman (1983):

When describing the attributes of a person, subjects consistently categorized "Bank teller and feminist" significantly higher than "Bank teller." Perhaps especially startling is that statistical knowledge had no effect on the outcome. The above item was provided to three groups: The first was a statistically naive group of undergraduates; the second, a group of first-year graduate students who had taken one or more statistics courses; and the third, a group of doctoral students in a decision science program who had taken several advance courses in probability and statistics (labeled statistically naive, intermediate knowledge, and statistically sophisticated respectively).

Most people perpetrated what is described as the *conjunction fallacy*: They thought the probability of the conjunction is higher than the probability of the elementary event. This implies, given the card example above, the probability of drawing a diamond picture card is greater than the chance of drawing a simple diamond. Here again, it is the stereotype of the feminist that leads us to this incorrect conclusion in decision-making.

The availability strategy

The availability strategy refers to estimates of frequency or probability that are made on the basis of how easily examples come to mind. Individuals judge frequency by assessing whether relevant examples can be easily retrieved from memory or whether this memory retrieval requires great effort. Since availability is typically correlated with an objective frequency of an event, the use of this strategy usually leads to valid conclusions.

However, since several factors that strongly influence memory retrieval are not correlated, they can distort availability in a way that leads to erroneous conclusions.

A classic study (Tversky and Kahneman, 1973) defined the availability strategy. In this study, subjects were asked to consider two general categories of words; those with the letter *k* in the first position and those with the letter *k* in the third position. The subjects were then asked to estimate the relative proportions of each kind of word. The researchers discovered that individuals guess that about twice as many words have the letter *k* in the first position relative to those that have *k* in the third position. In point of fact, just about twice as many words have *k* in the third position.

The results indicate that we are very familiar with considering words in terms of their initial letters and rarely ruminate on words in terms of their third letters. Accordingly, retrieval of words beginning with *k* is much easier—many more are available—than words with *k* in the third position. The availability and effortlessness of retrieval lead to the illusory conclusion that many more begin with *k*. Given that availability is very much associated with retrieval, it is highly influenced by variables related to memory such as recency and familiarity.

Recency and availability

Since memory for occurrences usually depreciates with time, more recent episodes are recalled more accurately. Since recent items are more available, we judge them to be more probable than they really are.

Familiarity and availability

Familiarity can also augment availability and by this means warping our thinking. On the whole, events that are highly familiar lead us to overestimate their likelihood compared to events that are less familiar. For instance, important issues in the news (what's newsworthy, what's not) elucidates the persuasive influence of journalism. Murders get reported, but random acts of kindness often do not. Hence, it is simpler to overestimate deaths from murders compared to deaths from disease. Recency and familiarity, then, are likely to adulterate a strategy that, otherwise, works quite well.

Illusory correlation and availability

Another stimulus on availability is the illusory correlation. Correlation refers to a relation between two variables (i.e., variables that co-vary).

In a positive relationship, when one variable rises, the other also does, however, in a negative relationship the variance is in opposite directions (when one increases, the other decreases). An illusory correlation refers to the belief that a correlation occurs when it does not. Illusory correlation is recognized to be the underpinning of many stereotypes and prejudices (e.g., specific racial, ethnic groups, gender, or body types are lazy, dumb, violent, unemotional, overly emotional, have rhythm, etc.).

Illusory correlations may happen as a result of the selectivity of information we embrace pertaining to other individuals. We remember the information that sanctions our stereotype and close our eyes to disconfirming information (i.e., typically as an exception to the rule). In learning prejudices from parents and peers, they offer us with a frame or schema for assimilating information from our social experience. The schema guides us to focus on the attributes of others that endorse the stereotype.

The simulation strategy and availability

A particular case of availability is called simulation strategy. That is, there are occurrences when we need to make decision choices in the absence of previous experience. The simulation strategy denotes the ease with which we can envision a situation or event.

Influences on representative strategy include sample size, base rate, and conjunctive fallacy

Sample size

Individuals try to include probability and statistics into their thinking. However, our decision-making involves small sample sizes.

Base rate

We usually ignore base-rate information and base our decisions on representativeness.

Conjunctive fallacy

The probability of the conjunction is greater than the probability of the basic event.

Influences on availability include recency, familiarity, illusory correlation and simulation strategies

Recency

More recent events are recalled more accurately.

Familiarity influences

Events that are highly familiar lead us to overestimate their likelihood relative to events that are less familiar.

Illusory correlation

Refers to the belief that a correlation exists when it, in fact, does not.

Simulation strategy

There are occasions when we need to make decisions in the absence of prior experience.

Individuals' strategies or heuristics can creep into the programming of algorithms. The aforementioned biases such as "confirmation bias" (when an individual acknowledges a result since it confirms a previous belief) or "availability bias" (placing higher emphasis on information relevant to an individual than equally important information of less familiarity) can render the interpretation of Artificial Intelligence engaged machine learning data pointless.

Six Dominant Algorithmic Pathways

A typical example of an algorithm that we use in our daily lives is a recipe. This set of instructions provides for us the ingredients we will need and directions on what to do with those ingredients.

However, what if you cannot locate the measuring cup. An algorithm can assist in locating the cup. In addition, an algorithm may be required for how to use a measuring cup.

An algorithm is a set of instructions as well as taking into account who or what is going to interpret those instructions. For example, if you provide directions to an associate detailing how to get from your business office to the next meeting place, your associate will only know how to get to the meeting place if s/he knows where your business office is located. The person is not capable (yet) of finding that particular meeting place from say, another location.

This is how an algorithm can be both simple and complex. In terms of computer algorithms, understanding what a computer is capable of doing is a fundamental part of formulating algorithms.

The downside of using an algorithm to solve the problem is that this process tends to be very time consuming. Therefore, facing a situation where a decision needs to be made very quickly, you might be better off using a different problem-solving strategy.

For example, a physician making a decision about how to treat a patient could use an algorithm approach, yet this would be very time-consuming, and treatment needs to be done quickly. In this instance, the doctor would instead rely on their expertise and past experiences to very quickly choose what they feel is the right treatment approach.

How sorting algorithms evolved

One of the first algorithms produced was the bubble sort routine. Bubble sort is a technique for organizing numbers, letters or words by looping through a data set, associating each set of values side-by-side, and exchanging them when necessitated for use.

This loop is repetitive until the algorithm can move through the entire list without requiring exchanging something, which implies the values are arranged correctly. This kind of algorithm is often denoted to as a recursive algorithm because it loops on itself over and over until it concludes the task.

This kind of algorithm might look as simple as:

1. Go to the first value.
2. Check that value against the next value and exchange positions if needed.
3. Go to the next value and reiterate the comparison.
4. If we are at the end of the list, go back to the top if any value was exchanged at some stage in the loop.

But bubble sort did not turn out to be the most *efficient* way of sorting values. As with time computers became more adept at doing complicated chores speedily, new sorting algorithms arrived for use.

One such algorithm scans through the first list and generates a second list of arranged values. This technique only makes a solo pass through the initial list, and with each value, it will loop through the second list until it locates the accurate place to put the value. Typically, it is more efficient than implementing the bubble sort method. This is where algorithms can become interesting or peculiar.

The bubble sort technique in many ways is destined to become one of the most inefficient methods of sorting values. Nonetheless, if the original list is pre-sorted properly, bubble sort can be one of the *most* efficient. The reason being is that the bubble sort algorithm will go through the list a single time and ascertain it is precisely sorted.

Unfortunately, we do not always know if our list is pre-sorted or not, so we have to choose an algorithm that is going to be the most efficient to use on an average across a large number of lists.

The definition of an algorithm is straightforward. It is a set of instructions. For example, an algorithm's instructions can be as follows:

1. Go to the first street light.
2. Take the first left.

3. Find the third house on the right.
4. Ring the doorbell.
5. Deliver the package.

Even though the definition of algorithm is straightforward, the actual meaning and how it affects our lives can be quite complex.

The Throughput Model provides six dominant algorithms that allows data input in a specific form. Further, it can assist in the presentation and form of the data whereby an algorithm can assist in discovering new patterns that were not apparent before viewing from the Throughput Modeling's perspective.

Decision makers' cognitive biases influence Artificial Intelligence through data, algorithms and interaction. Machine learning, a subset of Artificial Intelligence, is the capability for computers to be trained without explicit programming. Artificial Intelligence's training or learning is fashioned by data, algorithms, and experience through interactions and iterations. The size, structure, collection methodology, and sources of data impact machine learning. Machine learning is dependent on the quality of learning data sets. Similar to people, in Artificial Intelligence the more objective the data and the larger the data set, the less possibility of distortion.

When different kinds of human mistakes become baked-in parts of an Artificial Intelligence system could spell trouble. That is, our bias could be responsible for the selection of a training rule that fashions the creation of a machine learning model. Therefore, an Artificial Intelligence model will not be developed; but our own flawed observations inside of a black box. Therefore, the next discussion centers on six dominant algorithmic pathways that can, at least, alert us to the programming of algorithms.

There are six dominant pathways (Rodgers, 2006; Rodgers and McFarlin, 2017) that decision makers attempt to implement in order to reach a decision choice. The three primary and three secondary algorithmic pathways derived from the Throughput Model are shown in Figure 3.1.

I. Primary algorithmic pathways

1. The Expedient Pathway
2. The Ruling Guide Pathway
3. The Analytical Pathway

II. Secondary algorithmic pathways

4. The Revisionist Pathway
5. The Value-Driven Pathway
6. The Global Perspective Pathway

The *first algorithmic pathway* is the $P \rightarrow D$, the expedient pathway, which generally arises in settings where a decision choice should be

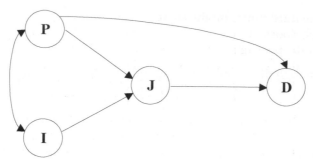

Figure 3.1. Throughput modelling process (Rodgers, 1997; Foss and Rodgers, 2011). Where **P** = perception, **I** = information, **J** = judgment, and **D** = decision choice. Single arrows from one construct to another indicate the hypothesized causal relationships explaining differences in decision outcomes. The double-ended arrow shows an interactive relationship

made promptly. The *second algorithmic pathway* is the **P→J→D**, the ruling guide pathway, in which time pressures may be domineering but are not as demanding as the **P→D** pathway. For the **P→J→D** pathway, a decision maker constructs the problem, evaluates it, and then makes a decision choice. The *third algorithmic pathway* is **I→J→D**, labelled as the analytical pathway whereby relevant and reliable information is the guarantee of beneficial decision choices. When engaging this pathway, information will directly impact on the judgment stage before a decision choice is made. Judiciously, the information is pre-determined and is weighted by other sources without biases (Rodgers and McFarlin, 2017).

The three higher level pathways are:

1. **I→P→D** which represents the *relativist-based algorithmic pathway*. This pathway updates knowledge or information by showing a pathway from information to perception. Further, this algorithmic pathway indicates that adequate information set can assist in revising a person or artificial intelligence system's previous manner of framing or viewing a particular problem before making a decision. In this particular pathway, a detailed analysis (judgment function) is ignored when making a decision. This pathway can handle a degree of time pressure since a detailed analysis in the judgment function is not required for decision-making purposes. In addition, the revisionist pathway is suitable for environmental changes due to this pathway allowing information to update or change an individual's perception function before making a decision. Therefore, the updated information people collect from their group to formulate their perception to reach a decision.

2. **P→I→J→D** reflects the *value-driven algorithmic pathway*. This algorithm suggests how an individual or artificial intelligence

system's perceptual framing helps guide and select certain types of information used in the judgmental function. This pathway is influenced by information processing limitations, complexity, and coherence between perception and the available information. Hence, to take this pathway, you allow your perception to modify and select the information that will be analyzed for a decision.

3. Finally, the $I \rightarrow P \rightarrow J \rightarrow D$ underscores the *global perspective algorithmic pathway*. The global perspective-based pathway is an approach that focuses on a larger set of information (e.g., big data), which can influence a person or artificial system rule-based algorithms. This pathway places special attention on the reasons for performing certain actions. Since information (I) moderates the rule-based pathway $(P \rightarrow J \rightarrow D)$, following a specific algorithmic rule is often not sufficient; instead, correct motivation will be needed which will be mostly determined by information (Rodgers, 2006; Rodgers and McFarlin, 2017).

These six algorithmic pathways are described as follows:

1. The Expedient Pathway $P \rightarrow D$
2. The Ruling Guide Pathway $P \rightarrow J \rightarrow D$
3. The Analytical Pathway $I \rightarrow J \rightarrow D$
4. The Revisionist Pathway $I \rightarrow P \rightarrow D$
5. The Value Driven Pathway $P \rightarrow I \rightarrow J \rightarrow D$
6. The Global Perspective Pathway $I \rightarrow P \rightarrow J \rightarrow D$

Information Drives Implementing a Particular Algorithm

Regrettably, the big data analysis for algorithmic decision-making processes does not guide us to rate incomplete or noisy information, inadequate understanding, and undifferentiated alternatives (Rodgers, 2006). For example, it does not guide us regarding the rating of alternatives when the outcome on a particular event is uncertain. Moreover, it has not provided rules for determining the optimal alternative under uncertainty. For example, we are inclined to identify important attributes or properties from an information set. In order to use an algorithmic analytical and programmatic decision-making process for a particular set of information, the concepts of *information, events,* and *uncertainty* must be acknowledged and discriminated for usage. This information set can be viewed as precise and/or vague (Rodgers, 2006).

Information is precise if it can be qualified of being interpreted in exactly one way. Information is vague if it is not unambiguously defined or cannot be understood in at least one precise way. Moreover, *events* or objects are vague if they cannot be completely ordered. For example, if we cannot order Product A, Product B and Product C from best to worst items for consumption, then these objects are vague.

Uncertainty can be also regarded as precise or vague. For example, let us consider rolling a dice and calculating the probability that the number 5 will appear on top. A "dice" (or die) is a small cube—a square, box-shaped solid with a different number of spots on each of its six sides (Rodgers, 2006).

The probability that a rolled dice will exhibit the number 5 is a *precise* uncertainty of 1/6. In addition, this is an exemplification of a *"problem-solving"* result when we can calculate an exact probability of occurrence. Unfortunately, in most everyday problems the precise uncertainty for each outcome is not known. Estimating the probability of rainfall two years from now in Chicago, Illinois, on four days in July requires *vague* uncertainties. Most financial, marketing, and investment problems are of this kind and the outcomes are catalogued as *"decision-making"* problems.

Considering the eight cells of Table 3.2, we can encapsulate the concepts of event, uncertainty, and information. Cell 2, whereby the precise uncertainty regarding precise events is represented by vague information, might occur in determining that Amazon's stock price will outperform a newly started internet firm that engages in the retail sale of consumer products and subscriptions, but without a profit history. The detailed information may be vague regarding management and the financial statements of the two firms if they are not completely available to us. Nonetheless, we may still be able to establish the preciseness of uncertainty and the event grounded upon our knowledge pertaining to Amazon Corporation and the last 20-year history of startup internet firms' stock prices.

Cells 1 and 3 include both types of uncertainty regarding precise events. These cells depict the traditional probabilistic approach of an analytical and programmatic decision model. From an investment banker's perspective, however, cell 4 is more appropriate when information (vague) is gathered from unaudited or incomplete financial statements and it is difficult to assess the probability of whether the stock price will double for a company in the next year (precise event).

Table 3.2. Uncertainty x event type x information presented precise uncertainty vague uncertainty

	Precise Information	Vague Information	Precise Information	Vague Information
Precise events	1	2	3	4
Vague events	5	6	7	8

(from Rodgers, 2006).

Under the analytical and programmatic pathway, cells 5, 6, 7, and 8 can never occur because the events must be precise to determine their rank, order, etc. For example, uncertainty about a vague accounting event pertaining to what month a company will distribute dividend checks can never be precise when compared to three other companies without a history of distributing dividend checks. Ordering these companies in terms of dividend payments is very difficult.

In other words, it is incorrect to apply an analytical and programmatic pathway decision model to vaguely defined events. Hence, an analytical and programmatic approach can at best account for only 50% of all possible conditions, as represented in Table 3.2. Therefore, it is necessary to use a model that incorporates and addresses all the conditions in Table 3.2.

Since an analytical and programmatic approach can only account for (at best) 50% of all possible situations, it is necessary to expand this model to capture the other 50%.

In addition, when information is viewed as vague it may not be relevant and reliable. For example, if a situation calls for exact numbers in order to open a combination lock, then vague or inexact numbers may not be sufficient. Therefore, if information has to be exact, then only 2/8 (25%) of the cells are useful. Recall that cells 5 and 7 (precise information) are not reliable and relevant since the events are vague. That is, if the events are vague then information loses it differentiating quality. For example, we may not be able to differentiate whether a Brand AAA refrigerator is better or worse than a Brand ZZZ refrigerator.

Throughput Modelling provides a framework for algorithms employed in the course of information gathering and analysis before arriving at a decision choice. Although the Throughput Model has organization, it also allows enough flexibility for users to recognize the arrangement of the four major functions (that is, perception, information, judgment, and decision choice) in viewing algorithms. Moreover, the Throughput Model depicts the interactions of four major concepts of decision-making and problem-solving. These four major concepts are perception, information, judgment, and decision choice.

Figure 3.1 illustrates how the process of an individual or organization's decision choice is made. If perceived causality is an important determinant of how an individual represents a problem and uses the information, it is necessary to know what determines the perception of causality in *process thinking*. Since decision-makers typically process information subjectively, it is interdependent with perception in the model. The situational information and the decision-maker's prior expectations or beliefs about the information are relevant in perceiving the degree of coherence between them. The interdependency and redundancy of perceptual effects and presented information have important effects on the kinds of judgment and strategies that individuals use.

In the Throughput Model, information also affects judgment. For example, information that is stored and retrieved from memory affects decision-makers' evaluations of investment or credit portfolios. Typically, before an individual can make a decision, that individual encodes the information and develops a representation for the problem. Finally, perception and judgment can affect decision choice. Errors, heuristics, biases, and context-dependent strategies may result from decision-making procedures of which decision makers are largely unaware, and these may have a direct impact on decision choice. The strategies of judgment that influence decision choice are under an individual's deliberate control.

Conclusions

The Throughput Model highlights three primary ways everyone comes to a decision, which are represented through algorithms. An algorithm is a process used in calculations and problem-solving operations. The three basic algorithms can then be further expanded making six algorithms for the way a decision can be made. The three secondary algorithms are very similar to the primary three algorithms just with additional concepts of either perception or information added to the algorithm. The concepts are all the same. Throughout the algorithm they are just placed in different orders of influence, which causes the outcome to be different.

These six algorithmic pathways can be based on handcrafted systems that use simple scoring mechanisms, or the identification of keywords, or natural language extraction. Rules may be articulated directly by programmers or be dynamic and flexible based on machine learning of contextual data. In addition, these six dominant algorithmic pathways include human influences in decision-making, which can entail setting criteria choices and optimization functions. Software engineering of algorithms will need to consider ways that support feedback and logging mechanisms, to allow for greater accountability in these six algorithmic pathways.

Throughput Modeling algorithmic pathways are very helpful in that a problem can be solved in multiple formula ways. This process helps in dealing with situations requiring a great deal of time. In addition, algorithms can be compared in order to determine which one works best in a given situation.

Using algorithms for problem solving of various types has a very long history and was not achieved overnight. In short, Throughput Modeling algorithmic pathways present a method of solving formulas due to the increasing availability of data. Further, the accessibility of computers with significant processing power has influenced the development of new algorithms to help predict behavior and automate decisions. This has opened up many novel applications with the aim of increasing productivity, and enabling more efficient and informed decision-making in government, businesses and public bodies whereby the Throughput Modeling algorithmic pathways can assist.

This is how an algorithm develops and go forward to a better design. Hence, an algorithm does not have to be run by a computer to be an algorithm. An algorithm is a process, and processes are all around us.

In psychology, algorithms are frequently contrasted with heuristics. A heuristic is a mental shortcut that permits individuals to swiftly make judgments and arrived at decision choices. These mental shortcuts are usually enlightened by our previous experiences and permit people to act quickly. Nonetheless, heuristics are really more of a rule-of-thumb. Furthermore, they do not always guarantee a correct solution.

The benefit of using an algorithm to solve a problem or make a decision choice is that it yields the best possible answer every time. This is useful in situations when accuracy is critical or where similar problems need to be frequently solved. In many cases, computer programs can be designed to speed up this process. Data then needs to be placed in the system so that the algorithm can be executed to come up with the correct solution.

Such step-by-step approaches can be useful in situations where each decision must be made following the same process and where accuracy is critical. Because the process follows a prescribed procedure, you can be sure that you will reach the correct answer each time.

This chapter presented six dominant algorithmic pathways that can be used for problem-solving or decision-making purposes.

These six algorithms are depicted as follows:

I. Primary algorithmic pathways

1. The Expedient Pathway: $P \rightarrow D$
2. The Ruling Guide Pathway: $P \rightarrow J \rightarrow D$
3. The Analytical Pathway: $I \rightarrow J \rightarrow D$

II. Secondary algorithmic pathways

4. The Revisionist Pathway: $I \rightarrow P \rightarrow D$
5. The Value-Driven Pathway: $P \rightarrow I \rightarrow J \rightarrow D$
6. The Global Perspective Pathway: $I \rightarrow P \rightarrow J \rightarrow D$

The usability of these algorithms is a function of the availability of reliable and relevant information as well as for accuracy or speed. If complete accuracy is required, it is best to use an algorithm. By using an algorithm, accuracy is increased, and potential mistakes are minimized. On the other hand, if time is an issue, then it may be best to use a heuristic. Mistakes may occur, but this approach allows for speedy decisions when time is of the essence.

Heuristics are more typically implemented in everyday circumstances, such as considering the best route to get from point A to point B. In general, a person's best option would be to utilize a route that worked well in the past. Nonetheless, computer-based algorithms allow us to map out every possible route and determine which one would be the quickest.

In sum, algorithms evolve over time as we find more efficient ways to do things and/or computers become more capable of carrying out complex tasks. Next, a particular type of algorithmic pathway is selected because they are "more" efficient "most" of the time for problem solving or decision-making purposes. Finally, just because an algorithm is more efficient most of the time does not imply that it is the most ethical algorithm to employ in a particular situation.

Algorithms are at work assisting people every day. Searching the Internet is aided by algorithms, which is attempting to find the best search results. For example, smartphones software can offer directions, whereby an algorithm decides the best route for an individual to take. Understanding the Throughput Model six algorithmic pathways can assist individuals regarding the basics of how decision choices are computed and finalized.

References

Kahneman, D. and Tversky, A. 1973. On the psychology of prediction. Psychological Review, 80: 237–251.

Newell, A. and Simon, H.A. 1972. Human Problem Solving. Englewood Cliffs, NJ: Prentice-Hall.

Pomerol, J-C. 1997. Artificial intelligence and human decision making. European Journal of Operational Research, 99: 3–25.

Rodgers, W. 1991a. Evaluating accounting information with causal models: Classification of methods and implications for accounting research. Journal of Accounting Literature, 10: 151–180.

Rodgers, W. 1991b. How do loan officers make their decisions about credit risks? A study of parallel distributed processing. Journal of Economic Psychology, 12: 243–265.

Rodgers, W. 1997. Throughput Modeling: Financial Information Used by Decision Makers. Greenwich, CT: JAI Press.

Rodgers, W. 2006. Process Thinking: Six Pathways to Successful Decision Making. NY: iUniverse, Inc.

Rodgers, W. 2019. Trust Throughput Modeling Pathways. Hauppauge, NY: Nova Publication.

Simon, H.A. 1969. The Sciences of the Artificial. Cambridge MA: MIT Press.

Tversky, A. and Kahneman, D. 1974. Judgment under uncertainty: Heuristics and biases. Science, 185: 1124–1131.

4

Survey of Biometric Tools and Big Data

"Data scientists realize that they face technical limitations, but they don't allow that to bog down their search for novel solutions. As they make discoveries, they communicate what they've learned and suggest its implications for new business directions. Often they are creative in displaying information visually and making the patterns they find clear and compelling. They advise executives and product managers on the implications of the data for products, processes, and decisions."

—Harvard Business Review

"If implemented responsibly, Artificial Intelligence can benefit society. However, as is the case with most emerging technology, there is a real risk that commercial and state use has a detrimental impact on human rights."

—Apple CEO Tim Cook

The application of developing technologies such as biometrics, Artificial Intelligence, machine learning, neural networks, and deep learning are offering the level of accuracy and security required in the identity authentication process to support the sharing economy and other digital, app-driven organizations.

Biometrics can be categorized as the automated recognition of people based upon their behavioral or biological features. The primary enhancement of physiological biometrics is permanence. That is, most of the elements it represents are stable and do not change over time. Irises, for example, do not change. Neither do the unique, scannable patterns of a person fingerprints.

In addition, research suggests that customary undertakings such as the way we speak, write or type are controlled by a pattern that can be just as unique. Therefore, behavioral biometrics are indubitably 'more secure' as these traits are notably difficult to steal or replicate. Several conventional examples implemented by commercial applications encompasses, signature recognition, mouth movement analysis and typing rhythm, which can be expanded to most common spelling mistakes people make.

Biometrics is becoming more instrumental in inaugurating ethics and trust in the sharing economy since biometric data is continuously being fed into Artificial Intelligence applications and they are becoming better at clarifying a person is who s/he claims to be predicated on collected fingerprints, iris, facial, voice, and behavioral data.

Moreover, the developments in data storage technology, high-resolution imaging, and big data analytics have made biometrics a viable option for organizations studying for novel ways to manage risk, engage with customers, and improve their customers' experiences. The application of emerging tools such as biometrics, Artificial Intelligence-related enhancements such as machine learning, neural networks, deep learning are providing the level of accuracy and security necessary in the authentication/identification process to energize the global economy and other digital, app driven organizations.

In addition, the implementation of biometric systems hinge on the straightforwardness with which individuals can utilize them. In systems design it is essential to contemplate training in the exercise of the system, ease of use (e.g., are numerous steps, problematic actions, or intricate procedures required?), and management of errors (e.g., how does the system recuperate from a mistake?). Conceiving practical systems also necessitate that the designers have some knowledge of the individuals' users and operators, the context in which they will utilize the system, and their motivations and expectations.

This chapter highlights (1) authentication and identification of biometrics tools, (2) big data, (3) datafication of our lives, (4) biometric tools and Artificial Intelligence, (5) physical biometrics and (6) behavioral solutions, privacy issues, and warehouse biometrics examples.

Biometric Authentication and Identification

One question to pose is how do biometric software systems know indeterminately that individuals are who they say they are? The system can compare biometric features (e.g., Google or Facebook) in a selfie with that of a government issued ID. An ID by itself is not sufficient to verify an identity. That is, it requires corroboration with biometrics and Artificial Intelligence to ensure the individual presenting the ID is in fact the person pictured.

In addition, Artificial Intelligence is being trained by trainers to verify IDs and other documents at a global scale—some organizations are training their algorithms with data pulled from more than 200 countries and are supervising more than 300,000 verifications daily. The databases are very sizable, and the algorithms are growing to be more intelligent daily (https://insidebigdata.com/2019/04/11/how-big-data-ai-and-biometrics-are-building-trust-in-the-sharing-economy/).

This implies that it may become quite impossible to counterfeit a digital identity with an organization adopting these innovative tools since users are compelled to take a selfie and then that is matched with a government issued ID. The Artificial Intelligence can identify fraud in multiple ways. For example, if the biometrics do not align, the government issued ID is fraudulent, and so on.

Biometric authentication (i.e., verification) is the process of comparing data for the person's characteristics to that person's biometric "template" in order to determine resemblance. The reference model is first store in a database or a secure portable element like a smart card. The data stored is then compared to the person's biometric data to be authenticated. Here it is the person's identity, which is being verified.

In this mode, the question being posed is: *"Are you indeed Mr or Ms XYZ?"*

Biometric identification entails establishing the identity of a person. The aim is to capture an item of biometric data from this person. It can be a photo of their face, a record of their voice, or an image of their fingerprint. This data is then compared to the biometric data of several other persons kept in a database.

In this mode, the question is a simple one: *"Who are you?"*

The proliferation of these new collection devices, a range of new developments in recognition technologies, innovative approaches to sensor and data fusion, and the emergence of powerful analytic tools provide organizations with valuable new capabilities to enhance marketing efforts and other operations. Artificial Intelligence professionals are working on the ability to merge biometric information with intelligence artifacts and products from the non-biometric environment. Add to this the emergence of new "big data" tools, technologies and capabilities for analyzing large-scale data inputs from multiple sources, and biometric-enabled intelligence promises to provide a powerful new means of analyzing and predicting organizations' performance and value (Rodgers, 2016).

Biometrics is the procedure of employing a digital representation of a person's physical or behavioral characteristics as a means to identify or verify that particular individual (Rodgers, 2010, 2012). Those physical characteristics can range from the simple (e.g., fingerprints) to the slightly or significantly more complex (e.g., voice scans, iris scans, or face recognition).

Generally, biometric analysis systems have a huge demand for a large amount of biometric data. This data must be stored and somehow protected from unauthorized access. These systems are dependent upon complex algorithms that sort data in ways that will achieve an identifying result in a given application. Developers implement vital features that are unique from one person to another in order to make biometric identification effective.

At present, biometric systems can be implemented in at least three distinctive manners:

1. *Authentication (i.e., verification)*: Ensuring that an individual is who s/he claims to be;
2. *Identification*: Ascertaining who an individual is (e.g., identifying a person in a crowd); and,
3. *Screening*: Establishing whether a person belongs to a 'watch list' of identities.

Biometrics can be employed as *'supervised'* systems, such as at border crossings and as part of immigration control, or as 'unattended' systems that are implemented remotely 'on-the-go' progressively more by means of sensors on mobile phones. With *supervised* systems, the environmental surroundings (e.g., lighting, ambient conditions) are structured and there is the prospect for people intervention should a problem arise. For example, if the ePassport gates at Heathrow Airport, which rely on facial recognition, are not operating, a passenger is still able to have his or her passport confirmed manually by a Border Agency official. The ePassport gates are also monitored by officials to inhibit the system being 'spoofed'.

Big Data

A key difference between "big data" and "Artificial Intelligence" is that big data is the raw input that requires to be cleaned, structured and integrated before it becomes useful. On the other hand, Artificial Intelligence is the output, whereby the intelligence results from the processed data.

"Big data" massive data collections that the global community is contributing to every day, is only getting larger. Within that data, if we know how to unlock it, lies the potential to construct astonishing new organizations and solve some of the world's biggest challenges.

Data is the fuel that powers Artificial Intelligence, and large data sets make it possible for machine learning applications to learn independently and rapidly. The great quantity of data we collect enables our Artificial Intelligence systems with the examples they need to identify differences, increase their pattern recognition capabilities, and see the fine details within the patterns.

"Big data" is an area that captures the manner in which to analyze, systematically extract information from, or otherwise deal with data sets that are too huge or complex to be dealt with by conventional data-processing application software. Data with many cases (rows) provide larger statistical power, while data with higher complexity (more attributes or columns) may steer into a higher false discovery rate (Breur, 2016). Big data challenges embrace capturing data, data storage, data analysis, search, sharing, transfer, visualization, querying, updating, information privacy and data source. In the beginning, big data was connected with three essential concepts: *volume, variety*, and *velocity* (Laney, 2001). Other concepts later attributed to big data are *veracity (i.e., how much noise is in the data)* and *value* (Goes, 2014).

Artificial Intelligence enables us to make sense of massive data sets, as well as unstructured data that doesn't fit neatly into database rows and columns. Artificial Intelligence assists organizations to develop new insights from data that was formerly locked away in emails, presentations, videos, and images.

Databases are becoming all the time more versatile and powerful. Additionally, to traditional relational databases, there exists powerful graph databases that are more proficient of connecting data points and detecting relationships, as well as databases that focus in document management.

Big data might recognize more than you realize. It signifies that the field of cognitive psychology can cleverly uncover things they never before thought possible. We are now firmly in the era of *big data*, whereby computers are encapsulating and processing the aspects of everything we do with all our interconnected devices in real time. Organizations view this as the Holy Grail for being able to predict whom, where, and when users will select their existing solutions, and what their future solutions must look like to be attractive. By some estimates, people now employ the same amount of data in any two days than in all of mankind history prior to 2000 (https://www.forbes.com/sites/martinzwilling/2015/03/24/what-can-big-data-ever-tell-us-about-human-behavior/#5952c9e461f9).

Generated "big data" is generated daily and encompasses the global community. Every digital process and social media exchange produce more data. Systems, sensors and mobile devices transmit it. Big data is arriving from multiple sources at an alarming velocity, volume and variety. As defined, big data is a large volume of structured and unstructured data that an organization utilizes in its day-to-day operations. Nonetheless, the amount of data is not the entire story. What is critical is the manner in which organizations deal with this data for the benefit of their operations. This concept gained thrust in the early 2000s when industry analyst Doug Laney articulated the now-mainstream definition of big data as the three (with two more added) Vs (https://www.gartner.com/en/documents/2057415) along with veracity and value (Goes, 2014):

1. *Velocity*—The streaming of data takes place in unprecedented speed; that is why it should be managed in a timely way. Using sensors, RFID tags and other tools can help deal with the flow of data in near real time. For example, social media messages going viral in seconds, the speed at which credit card transactions are reviewed for fraudulent activities, or the milliseconds it takes trading systems to analyze social media networks to select signals that trigger decisions to buy or sell shares. Big data technology affords the analysis of data while it is being generated, without ever putting it into databases.

2. *Volume*—Organizations collect data from distinct sources such as business transactions, social media, and other relevant data. For example, the emails, twitter messages, photos, video clips, sensor data, etc., produced and shared every second has exceeded terabytes into zettabytes and brontobytes.

3. *Variety* suggests that all data can be presented in a variety of formats—from structured numeric data to the unstructured ones, which comprise text documents, audio, video, and email. Most of the world's data is now unstructured. Hence, it cannot be simply placed into tables (e.g., photos, video sequences or social media updates). However, with big data technology, different kinds of data can be harnessed (structured and unstructured), which include messages, social media conversations, photos, sensor data, video or voice recordings. This type of data can be placed together with more traditional, financial structured data.

4. *Veracity* denotes the messiness or trustworthiness of the data. With many types of big data, quality and accuracy are less controllable (e.g., Twitter posts with hash tags, abbreviations, typos and colloquial speech as well as the reliability and accuracy of content). Nonetheless, big data and analytical tools can improve the data veracity. That is, volumes often times make up for the lack of quality or accuracy.

5. *Value* implies that organizations make an effort to collect and leverage big data for a purpose. Further an analysis of costs and benefits for the use of big data is necessary.

In terms of Artificial Intelligence algorithms gathering mechanisms, there are four major areas to collect data from: existing datasets, social media, smart phones, and wearable devices. This suggests that the use of reliable information can assist in the development of apps to understand and type people more efficiently. The University of Cambridge's Apply Magic Sauce leverages data to analyze social media pages in order to provide individuals psychological information about themselves or for that matter, a company or researching body about a group of people. Likewise, IBM's Personality Insights can inform you about yourself by analyzing a section of text.

These algorithms take into account much more than just status updates and "about me" texts. They also examine friends' lists, draw correlations, and see density. That is, how strong a real social network really is in place. They view information about when the user actually connected with the world of social media, how often they post, and pages fond of or apps were implemented. These details convey specific information about an individual's personality.

Further, by mining through logs of data, researchers can create far better algorithms that may improve people's lives on a daily basis. In addition, that individual data may be gathered and used to create personalized algorithms. That is, when two people search for the same set of words, they will yield different results, due to the engine understanding the mind of the user. Another way of viewing this issue involves better combining of information. For example, if users search "basketball playoffs" in the late spring, they are likely referring to professional NBA basketball; searching in early spring, on the other hand, likely refers to college basketball. It is not, however, quite clear how useful these algorithms are for classifying and categorizing information. While search engines are constantly updating and optimizing, individuals may find themselves struggling to search for certain topics when the search engine naturally undertakes it knows what a user wants. Hence, the way forward is bigger, better data.

Big data implemented in research is also producing noteworthy novel information, and occasionally showing that old assumptions incorrect. For example, big data assisted Rosalind Franklin University of Medicine and Science researchers discredit various assumptions regarding the variations between male and female brains. Specifically, that the females have a bigger hippocampus that amalgamates new memories and links emotions to the senses. These researchers conducted their study by joining the findings among 76 published papers, covering over 6,000 test subjects. It seems bigger is better when it comes to testing pools. This underlines two of the major benefits big data provides for psychology research. That is, more efficient exploration of data, and cross-validated exploration of data. Hence, researchers can comprehend more, quicker, as well as observe beyond their own testing circle (see https://dataconomy.com/2016/04/big-data-tackles-personality-behavior-psychology/).

Invented by Tal Yarkoni, of the University of Texas, Neurosynth is a software framework for large-scale automated synthesis of functional neuroimaging data. In the same way, this program integrated over 9,000 neuroimaging studies. The data comprised 300,000 brain activations. The sizable amount of data assisted in steering researchers toward a subject swiftly and precisely. Their testing illustrated that the program operates nearly as well as manual research, which necessitates hundreds of hours of more work. Big data can assist researchers go through the information in a much improved manner that might otherwise take an eternity to scrutinize (http://pilab.psy.utexas.edu/publications/Yarkoni_NatureMethods_2011.pdf).

Marketing is one of the most recognizable and widespread uses of big data. Researchers and psychologists can utilize some of the same data and do something a little more impactful. Although commercial firms are still propelling big data to marketing and advertising purposes, those designs of psychological analysis will, and do, open doors for ordinary consumers, as well. For example, a smart phone can tell where you are and how active you are, and may provide more assistance in the future. Furthermore, most of that data can come together and let you know, at least objectively, how satisfied or well off you are. This technology may lead to improving self-understanding, and may be the driver for more fascinating new apps.

Datafication of Our Lives

Below are essential ways that Strong (2015) outlines how our lives are becoming increasingly more data led:

1. *Datafication of emotions and sentiment*: The eruption of self-reporting on social media has led people to offer the innermost confidential elements of ourselves. Several market research organizations now utilize this data by "scraping" the web in order to gather meticulous examples of the sentiment relating to certain issues, brands, products, and services.
2. *Datafication of relationships and interactions*: We are now not only able to see and track the manners that individuals relate, but also with whom they relate, how they do it, and when. Social media has the capacity to renovate our understanding of relationships by data-oriented professional and personal connections on a global scale.
3. *Datafication of speech*: Speech analytics is turning out to be more commonplace, mainly as conversations are all the time more and more recorded and stored as part of interactions with call centers, as well as with each other. As speech recognition expands, the range of voice-based data and connotation that can be portrayed in an intelligible format are flourishing.
4. *Datafication of offline and back-office activities*: Within many data-intensive provinces such as finance, healthcare, and e-commerce, there is a large amount of data stored on people behaviors and outcomes. In addition, the advent of image analysis and facial recognition systems processing in-store footage, traffic systems, and surveillance.
5. *Datafication of culture*: There is a whole new discipline of "cultural analytics" that incorporates digital image processing and visualization for the analysis of image and video collections to explore cultural trends. For example, Google's Ngram service has already datafied

over 5.2 million books from 1800 to 2000 to let anyone analyze cultural trends (Strong, 2015).

Major challenges linked with people cognitive processing of big data

Some of the major challenges associated with individuals' cognitive processing of big data, and using it effectively, includes the following (Strong, 2015):

(a) *The individuals' psychology of cognitive inertia*: People appear to be wired to resist change, with a set of cognitive 'rules of thumb' (i.e., heuristics and biases) that pertains to short-term loss-averse behaviors (Tversky and Kahneman, 1974). People are inclined to rely on familiar assumptions and display a reluctance to revise those assumptions (e.g., confirmation bias), even when new evidence contests their accuracy.

(b) *Cognitive ability to make sense of data*: Although computers can process and stockpile large volumes of data, gauging the implications is part of individuals' decision-making propensities (Rodgers, 2006). Decision-making propensities is the process of obtaining meaning from experience and situational awareness that appears to be a strain for both people and computers.

(c) *Information overload and data quality*: In reality, more data does not automatically lead to improved decision choices (Rodgers, 1993). More information typically implies more time is necessary in order to arrive at a decision choice. This process can lead to inertia, or volumes of one type of data biasing the decision choice in an unsuitable path, since more data is not continually superior data.

Unquestionably, driving towards more data connection online and offline, will explain more and more about people. This process enables improvement pertaining to projections of security matters, forecasting, purchasing habits and interests, etc. On the other hand, the challenge of anticipating what's in store necessities and behavior is much more problematic. For this reason, what is paramount is people's cognitive processes funneling big data in organizations, rather than the other way around, for the near future.

How big data is utilized in data analytics

The variety of kinds of data analytics are sometimes broken into four groupings of descriptive, diagnostic, predictive and prescriptive analytics (Table 4.1). Beginning from the left of the table, descriptive analytics examine historical data to answer questions; and diagnostic analytics

Table 4.1. Types of data analytics

Descriptive Analytics	Diagnostic Analytics	Predictive Analytics	Prescriptive Analytics
What happened?	Why did it happen?	What will happen?	What do we do?
Analyzes historical data to answer questions.	Recognizes patterns and discovers associations in data.	Implements historical and current data to predict future occurrences.	Employs rules and modeling for improved decision-making.

← Fundamental values and insights   Higher value and forward-thinking insight →

identifies patterns and discover relationships in data. Further, predictive analytics utilizes historical and current data in order to forecast future activity, while prescriptive analytics applies rules and modeling for enhanced decision-making activities.

While most organizations use an amalgamation of the aforementioned different types of analytics, in order to gain the following advantages:

a) A better understanding of current and future performance
b) Deeper knowledge of customer behaviors
c) Essential performance indicators (i.e., KPIs) for decision-making
d) Elevated capacity to experiment, learn, and improve, and
e) An exploitative advantage over lesser state-of-the-art participants.

Moreover, the exploitative advantage of advanced analytics can be enlarged by constructing an effective Artificial Intelligence capability. There is far more data being produced today than individuals can analyze in any significant manner. Tools such as machine learning, predictive analytics, and data visualization can assist in finding meaningful relationships by tunneling deeper into large data sets and enhancing the speed and preciseness of decision-making.

Leveraging Artificial Intelligence and advanced analytics in driving value creation and future growth can enhance an organization's success. Some practical applications of Artificial Intelligence and analytics embrace:

1. Tracking and forecasting relevant exponential technology trends. In addition, implementing data analytics along with Artificial Intelligence proactively can assist an organization in determining how and when to take action, to make better decisions, and stay ahead of competitors.
2. Using predictive analytics to reduce decision choices grounded on intuition or outdated models. In other words, relying upon the highest paid individual's opinion instead of relevant data.
3. The capability to benchmark and track the progress and speed of individual's innovative projects through development phases, and forecast future outcomes and revenues.

The significance of being able to call together around Artificial Intelligence and data analytics set of shared objectives and purpose can eliminate many mistakes by organizations. Most organizations are on a journey to the future, and they should take that Artificial Intelligence journey as a motivated group with a shared purpose.

Biometric Tools and Artificial Intelligence

Biometric solutions are typically used for prevention and/or detection for security and access control across organizations and government units. Businesses, non-profit organizations and governments have taken dedicated interest in Artificial Intelligence and biometric applications. Further, they have been aggressively funding advanced research programs that offer Artificial Intelligence-related biometrics.

Biometrics can be viewed as the authentication of individuals, it is what makes them who they are, it is the measure of someone's biological characteristics. Biometrics includes how the individual looks like, their facial features, their fingerprints or palm print. As technology is evolving biometrics is being used as an identification factor for many places where confidential or personal information is included. Fraud is defined as the criminal intent to take personal or confidential information for someone's own personal gain. In order to prevent fraud, biometrics are being used in the simplest items such as cellular devices to more complex things such as accessing a bank account or a safe. Personal devices use protection such a password, or a personal identification number (PIN), or even fingerprint recognition. Bank accounts use more complex protection such as facial or voice recognition to help prevent fraud with more complex and personal information. Using these protective authorizations with biometrics it is much harder for anyone to steal personal information and use it to their advantage. The decision-making model starts with information, without it no decision can be drawn therefore nothing will be accomplished. As technology advances the use of biometrics also advances. The biometric technologies can be separated into two categories—physiological and behavioral.

How Artificial Intelligence supported biometric technologies work

There are two categories of biometric identification and recognition solutions: *Physical* and *behavioral*.

Physical biometric solutions use distinctive and measurable characteristics of particular parts of the human body, such as a person's face, iris, DNA, vein, fingerprints, etc., and transform this information into a code understandable by the Artificial Intelligence system.

Behavioral biometric solutions operate in a similar way, except they use unique behavioral characteristics, such as a person's typing rhythm, way of interaction with devices, gait, voice, etc. This encoded biometric information is stored in a database and digitally sampled during authentication and verification.

Physical Biometric Solutions

Some examples of the physiological characteristics in biometrics include the use of fingerprint, palm print, hand geometry, iris, scan, retina scan, face recognition, DNA, and many more. These biometrics are used in everyday life with gadgets at home such as cellular devices to some not so common gadgets like accessing vaults or storage of valuables.

There are many other types of physiological characteristics in biometrics; however, those discussed next are the most common ones used today. Perhaps as Artificial Intelligence continues to advance, other types of biometrics will soon become more common and will be a part of someone's life on daily basis.

Facial recognition

A very popular biometric that is being used a lot more frequently now is facial recognition. It was not very popular because technology wasn't as advanced as it is now. Artificial Intelligence applied facial recognition is a lot more advanced and it provides for use of some of these techniques in a lot more user-friendly manner. The manner in which facial recognition operates begins with the use of a camera or a picture of an individual to recognize whether a positive identification exist.

If a device is equipped with a camera, it can easily be used for authentication. Such as facial recognition. A facial recognition biometric system identifies and confirms an individual by extracting and comparing selected facial features from a digital image or a video frame to a face database. For example, an algorithm may evaluate the distance between the eyes, the width of the nose, the depth of the eye sockets, the shape of the cheekbones, the length of the jaw line, etc., and put into code the corresponding data as "face prints," which can then be implemented to find suitable matches in a destination database.

This tool has advanced greatly because now an individual does not have to be right in front of the camera or close to the camera for that matter. An individual might be scanned from a distance and they will not even know it. It has also advanced greatly from simply matching someone's pattern to now matching distinctive features and complex points in someone's face. This was mainly used by officials before in missing case or as security measures in places like an airport. Now this

Figure 4.1. Facial recognition image

type of technology is used almost every day. Moreover, facial recognition is used in cellular devices, tablets and even computers, this comes to show how much technology has advanced in a short period of time.

DNA matching

DNA is a very well-known physiological biometric, it is the double helix structure presented in the human cells. DNA can be used to produce a DNA fingerprint or a DNA profile. DNA can be taken from many different sources, blood, hair, saliva, finger nails, and anything else that was once attached to the body at one point. The one downfall to DNA biometrics is that it is very slow, it cannot recognize someone within a matter of seconds or even minutes. Therefore, results will take weeks and months to be received and that can be very frustrating. Another downside of DNA is that it is very costly therefore it will not be used as frequently as other biometrics will be used (Figure 4.2).

This tool represents the identification of an individual using the analysis of segments from DNA. Today, DNA scans are used for the most part in law enforcement to identify suspects. In the past, DNA sequencing has been too slow for widespread utilization for commercial use. Nonetheless, DNA scanners prices are dropping and becoming more affordable. Moreover, the identification of all the building blocks in the human DNA have opened the door to a flood of new reports about genetic links to disease. That is, scientists are scanning human DNA with a precision and scope once unthinkable and rapidly finding genes linked to cancer, arthritis, diabetes and other diseases (https://www.seattletimes.com/business/dna-scanning-leads-to-breakthroughs/).

Note, that there are some exceptions to using DNA. For example, a person can have a condition called chimerism, which means s/he has

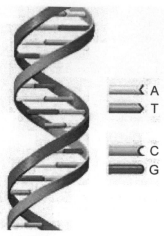

Figure 4.2. DNA image

two sets of DNA each with the genetic code to make a separate person. The rare condition can happen during fetal development. A chimera is a single organism that's made up of cells from two or more "individuals". It contains two sets of DNA, with the code to make two separate organisms. One way that chimeras can happen naturally in humans is that a fetus can absorb its twin. Further, given this condition, a mother given birth may not display the same DNA with the child (https://www.scientificamerican.com/article/3-human-chimeras-that-already-exist/).

Ear recognition

Represents the identification of an individual using the shape of the ear. A person's ear is a very good source of data for passive individual identification. Further, a person's ear appears to be a useful candidate solution since ears are visible, their images are easy to take, and the structure of an ear does not change radically over time (Figure 4.3).

Figure 4.3. Ear recognition image. http://www.isis.ecs.soton.ac.uk/images/newear.gif

Eyes—Retina and iris recognition

Represents the use of patterns of veins in the back of the eye to undertake recognition. The retina scan is not as common, but it is being used with very safe and confidential information. The way the retina scan works is that the individual is forced to look at a specific point and focus on that point, then the scanner can scan the individual's retina. The retina scan is not very user friendly but it is very accurate and can distinct one person from another (Figure 4.4).

The iris scan is also very similar. This method uses high resolution images to compare the patterns in someone's iris. This type of technology is used with a camera and will reduce the amount of light to therefore reduce the amount of reflection from the convex cornea to be able to make a very detailed picture. This process then uses the picture to make a conclusion as to the identification of that person.

Iris recognition is an automated method of biometric identification that utilizes mathematical pattern-recognition techniques on video images of one or both of the irises of a person's eyes, whose complex patterns are unique, stable, and can be seen from some distance (Figure 4.5).

Figure 4.4. Retina scan image. http://www.westtexasretina.com/images/normaleye_150.jpg

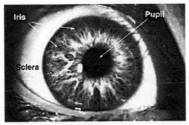

Figure 4.5. Iris, pupil, and sclera image. http://webvision.med.utah.edu/imageswv/pupil.jpeg

Finger geometry recognition

The use of 3D geometry of the finger to determine identity. Biometrics grounded on 3D finger geometry exploit discriminatory information provided by the 3D structure of the hand, and more precisely the fingers, as encapsulated by a 3D sensor. The advantages of current 3D finger biometrics over traditional 2D hand geometry authentication tools are enriched accuracy, the capability to work in contact-free mode, and the ability to merge with 3D face recognition using the same sensor (Figure 4.6).

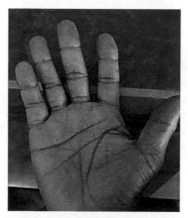

Figure 4.6. Finger and hand geometry

Hand geometry recognition

The use of the geometric features of the hand such as the lengths of fingers and the width of the hand to identify a person. The hand geometry can be taken as an image of your entire hand or as a two-finger reader only, and just like it sounds, an image of just two fingers will be taken. The devices to take a hand geometry are very accessible and small that it can be taken pretty much anywhere. As technology advances the goal is to be able to store all these palm prints and finger prints into one big database that is accessible to many people to try to figure out the identity of someone's fingerprint or palm print (Figure 4.6).

Body odor (Scent)

It is the use of an individual's odor to ascertain identity. Odor biometrics attempts to identify individuals based on a unique chemical pattern. Their applications cover from individual identification in airports, to the detection of different components in the human body. Body odor is implemented as a biometric identifier with high rates of accuracy, due to

Figure 4.7. Scent image. http://static.howstuffworks.com/gif/
istock_000002439257medium.jpg

chemical patterns in the smell that are unaffected by bodily changes or deodorant (Figure 4.7).

Fingerprint recognition

Most fingerprint biometric solutions scan for specific features of a fingerprint, such as the ridge line patterns on the finger, the valleys between the ridges, etc., generally referred to as minutiae, which are then transformed to warehoused digital data. In order to get a fingerprint match for verification or authorization, biometric systems must locate a sufficient number of minutiae patterns. This number varies across systems.

The use of fingerprint is the most common one and it can be used for identification and verification. It can be used to help identify who you are without any other sort of identification or it can also be used to verify who you are while comparing another form of identification. The fingerprint can be used by two different approaches, the minutiae-based approach and the pattern-based approach. The minutiae approach implements the ridge ending and bifurcation on the finger to plot points known as minutiae. This approach makes it easy to compare two distinct fingerprints in an electronic system. The second approach is the pattern-based approach and it is performed in two blocks called image enhancement and distortion removal. In both approaches the fingerprint is being scanned and can be used to make a reasonable conclusion on your verification or identity.

Moreover, fingerprint scanners have become ubiquitous in recent years due to their prevalent deployment on smartphones. Any device that can be touched, such as a phone screen, computer mouse or touchpad, or a

Figure 4.8. Fingerprint image. http://education.vetmed.vt.edu/curriculum/vm8054/labs/Lab14/IMAGES/FINGERPRINT.jpg

door panel, has the makings to become an easy and convenient fingerprint scanner.

Palm print works very similarly to fingerprint recognition. Every palm print is distinct, so it is also fairly easy to compare to others and make a decision to an individual's identity. The palm is also scanned and the ridge patterns, endings and paths of the raised portion of the palm are all measured.

Palm vein technology can authenticate individuals, especially when situations required highly responsible security and super-sensitivity. It is comparable to fingerprint analysis. A camera takes the photo and studies the palm prints and other physical traits to uniquely identify an individual's palm.

Vein (Finger) identification

This identification method uses an individuals' vein along their wrist or underneath the surface of their finger. It has been learnt that every person's vein pattern is unique and cannot be matched with others. Wrist or finger vein identification is a highly secure method as it is on their wrist or beneath the finger surfaces and thus cannot be destroyed or changed.

Facial thermo imaging

Face thermography is a form of biometric that is being used more frequently. This technique works by detecting heat patterns in someone's face by the shape of their blood vessels. There is no physical contact as a camera will import a picture and will be collected quickly. This type of biometric is very similar to facial recognition.

Thermal imaging is simply the procedure of using the heat given off by an object to yield an image of it or locate it. Thermal imaging frameworks for detection, segmentation and distinctive feature extraction and similarity measurements for individuals' physiological biometrics recognition specialized on algorithms that would extract vasculature information, create a thermal signature that identify the person. The highly accurate results attained by the algorithms demonstrate the capability of

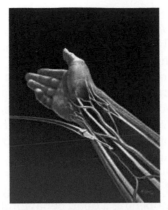

Figure 4.9. Vein patterns image. http://s3.images.com/huge.56.284299.JPG

Figure 4.10. Thermo imaging recognition image. http://biometrics.pbworks.com/f/facial%20thermography.png

the thermal infrared systems to extend in application to other thermal imaging-based systems, such as body thermo imaging.

Behavioral Biometrics

Different type of behavioral biometrics includes voice recognition, gait recognition, dynamic signature and keystroke dynamics. Behavioral identifiers are typically being implemented in conjunction with another method due to its lower reliability. Nevertheless, as technology improves, these behavioral identifiers are increasing in status. Unlike physical identifiers, which are restricted to a particular fixed set of individual characteristics, the only limits to behavioral identifiers is the human imagination. For example, this approach is often utilized to distinguish between a human and a robot. This process can aid an organization to filter out spam or detect attempts to brute-force a login and password. As technology advances, the systems are likely to improve at precisely identifying individuals, but less effective at distinguishing between humans and robots.

Behavioral biometrics identify and measure individual activities, such as keystroke dynamics, voice print, device usage, signature analysis, error patterns (e.g., accidentally hitting an "l" instead of a "k" on two out of every fifth transaction), etc. Such behavioral biometrics are usually implemented as a supplementary layer of security, along with other credential or biometric information.

Whereas, most physical biometric solutions systems authenticate the user only once and typically at the beginning of an action, such as logging into a device or opening a door. Behavioral biometric technology endeavors to plug the gap of authentication in a situation during an action.

For instance, the original user may offer her/his credentials to another individual subsequent to the user being successfully authenticated. In order to minimize such possibilities in this case, behavioral biometric solutions scrutinize users' interactions with their devices, recording activities that fluctuate from normal usage patterns.

Speaker (Voice) recognition

Voice-based digital assistants and telephone-based service portals are already using voice recognition to identify users and authenticate customers. Speaker or voice recognition diverges from speech recognition in that the former recognizes and identifies a speaker implementing voice biometrics and the latter scrutinizes what is being said. Voice biometrics comprises physical and behavioral characteristics. The physical features represent the shape of the vocal tract responsible for articulating and controlling speech production, whereas, the behavioral features includes pitch, cadence and tone, etc.

Voice biometric solutions consist of encoded frequencies or formats and produces a model "voice print" distinctive to an individual. This voice print is utilized for identification and authentication of the speaker. Moreover, in terms of collection, voice biometrics works by digitizing a profile of a person's speech to produce a stored model voice print, or template. Biometric technology decreases each spoken word to fragments constituted of several dominant frequencies named formants. Each fragment has several tones that can be encapsulated in a digital format. The tones jointly identify the speaker's unique voice print. Voice prints are stored in databases in a manner comparable to the warehousing of fingerprints or other biometric data.

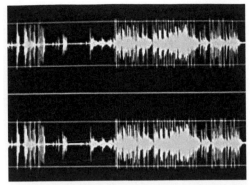

Figure 4.11. Voice recognition image. Speaker recognition Website: http://www.poegles. com/tag/science-daily/

In summary, there is a distinction between speaker recognition (acknowledging who is speaking) and speech recognition (comprehending what is being said). These two terms are oftentimes muddled, as is voice recognition. Voice recognition is a synonym for the speaker, and thus is not the same as speech recognition.

Voice recognition biometrics is the ability to recognize an individual's voice, just like it sounds. Speaker or voice recognition relies on the structure of a person's vocal tract and their personality. Speaker and speech recognition are not the same thing, speech recognition is the recognition of certain words not someone's voice. One big downside of voice recognition is that it will not recognize an individual on a recording but it has to be in person. The other downside is that it depends on a scripted text.

Moreover, there is a difference between the act of authentication (commonly referred to as speaker verification or speaker authentication) and identification. The coverage on natural language processing in Chapter 1 explored voice recognition in greater depth.

Keystroke recognition (Typing)

Keystroke dynamic biometrics is the way people type. This tool represents the use of the distinctive features of an individual's typing in order to determine identity. That is, everyone has a distinct typing style—the speed at which they type, the length of time it takes to go from one letter to another, the degree of impact on the keyboard. This biometric measures the different types of speed and the timing between every key pressed on the keyboard. Everyone takes more time typing out certain words or if it is something they are used to typing like a username or a password they might type faster. Therefore, the timing between keys or words is distinct and can be used to detect the identity of an individual.

Figure 4.12. Keystroke dynamics image. Website: http://blogs.technet.com/steve_lamb/archive/2006/03/13/421925.aspx

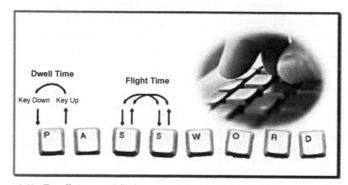

Figure 4.13. Dwell time and flight time. Website: http://articles.techrepublic.com.com/5100-10878_11-6150761.html

It is highly advantageous for lowering fraudulent activities such as improper emails and unethical typing since keystroke recognition measures an individual's typing patterns. Computers utilizing keystroke recognition software can identify the criminal activities and pushes the message to the right people for them to take proper action.

There are not as many behavioral biometrics currently being implemented as physiological biometrics. Nonetheless, as technology and artificial intelligence advance the use of these biometrics will also advance.

Signature recognition

Dynamic signature is the ability to recognize if someone has actually signed or even written a word using a pen or pencil and paper. The way every individual writes is distinct and the way everyone signs is even more distinct. Dynamic signature will analyze the various characteristics in someone's signing style to determine if it really was that person or

Figure 4.14. Dynamic signature image. Website:http://spie.org/x2434.xml?parentid=x2410 &parentname=Electronic%20Imaging%20&%20Signal%20Processing&highlight=x2410

someone else. The angle of the signature, the amount of pressure on certain letters, the formation of the letters and many other things can distinguish a person's signature.

Digital signature scanners are in wide use today at retail checkouts and in banks. The authentication of a person can be made through evaluation of handwriting style, particularly the signature. "Static" and "dynamic" are the two types of key for digital handwritten signature authentication. Static is oftentimes a visual comparison between one scanned signature and another. Moreover, it can represent a scanned signature against an ink signature. Technology is available to inspect two scanned signatures utilizing enhanced algorithms. This data can be employed in a court of law by using digital forensic investigative tools. Further, it can be crafted as a biometric template whereby dynamic signatures are authenticated either at the time of signing or post-signing. This technology can also be used as triggers in workflow processes.

Gait

This biometric depicts the manner in which someone walks, which is unique to an individual. Gait can be used to authenticate employees in a building, or as a secondary layer of authentication for particularly sensitive locations.

Gait recognition is also becoming popular these days as it has the ability to recognize the unique way someone walks. It analyzes a lot of factors like how fast someone is walking, the length of their legs and many others. It is also an example of a passive biometric, the individual does not have to do anything specific or agree to taken into consideration anything but just do the simple task of walking. Someone can fake their

Figure 4.15. Movements in a gait repeat as a person cycles between steps

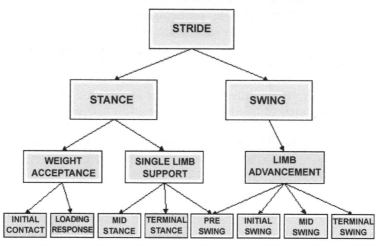

Figure 4.16. Gait recognition chart. http://sprojects.mmi.mcgill.ca/gait/normal/chart.gif

gait by purposely walking differently, taking shorter steps or longer steps, walking slower or faster. There are several ways to beat the system but if that individual knows about gait recognition, it will not try to beat the system.

Big Data Influencing Physical and Behavioral Features

Engagement patterns (iris, facial recognition, ear, hand geometry etc., gait etc., can be combined in an Artificial Intelligence system for detection and prevention purposes. Big data allows for interaction with technology in diverse ways. For example, navigation pattern can combine gait, keyboard dynamics and signature recognition can be used in combination to detect or prevent use entry or access into a room or building. In addition, mouse movements and finger movements on trackpads or touch-sensitive screens are unique to individuals and relatively easy to detect with software, no additional hardware required.

Dedicated cameras collecting data on face, ear, iris recognition to be capture and downloaded onto a database. The manner in which we look at people, places and things, open and use apps, how low we allow our battery to become, the locations and times of day we are most likely to utilize our devices depict unique physical and behavioral characteristics. Further, the manner we navigate websites, how we tilt our cellular devices when we hold them, or even how often we check our social media accounts are all potentially distinctive physical and behavioral characteristics. These behavior and physical patterns can be utilized to discriminate people from bots, until the bots get better at imitating humans. Moreover, they can also be implemented in amalgamation with other authentication methods, or, if the technology enhances enough, as standalone security measures. Finally, as a global community the issues dealing with ethical consideration are extremely important to take into account as depicted in Chapters 5 and 6.

Biometrics summary

As these types of biometrics are used for privacy and information protection purposes the level of fraud begins to increase. People will find ways to beat the system and pretending to be someone they are not to be able to get into certain databases. There are three different categories that tie into authentication factors, something you know, something you have, and lastly something you are. Something you know can be as simply as knowing someone's passwords, or username to something more complex like personal information. Security questions are now being used more frequently and if someone knows some personal information, they might be able to guess the answer to the security question. Something you have can be like a credit card or a debit card to be able to access the funds within that account. Something you are is being tied into biometrics with using certain features that only that individual has. These can be things like their fingerprint, voice recognition and many more. This type of features can be more protective but will not assure that information will not be stolen, therefore a combination of the three types should be used to help keep information safe.

For the authentication process, the goal is to verify the identity of a person. Ideal biometric information should have the following multiple properties:

- Uniqueness: this information must as dissimilar as possible for two different individuals;
- Universality: all individuals must be characterized by this information;
- Permanency: it should be present during the whole life of an individual;
- Measurability: it can be evaluated (in an easy way);
- Acceptability: it concerns the possibility of a genuine application by users.

Table 4.2. Overview of selected biometric technologies

Biometric	Uniqueness	Universality	Permanency	Measurability	Acceptability
Facial Recognition	Low	High	Medium	High	High
DNA Matching	High	High	High	Low	Low
Ear Recognition	Medium	Medium	Medium	Medium	Low
Eyes-Retina & Iris	High	High	High/ Medium	Medium/ Low	Low
Finger/Hand Geometry	Medium	Medium	Medium	High	Medium
Body Odor	Medium	Medium	Medium	Low	Low
Fingerprint	High	Medium	High	Medium	Medium
Finger Vein identification	High	Medium	High	High	High
Palm Vein authorization	High	Medium	High	Medium/ Low	Low
Facial Thermo Imaging	High	Medium	High	Medium	Low
Speaker (Voice) Recognition	Low	Medium	Low	Medium	High
Keystroke dynamics	Low	Medium	Medium	High	High
Signature	Low	Medium	Low	High	High
Gait	Medium	Medium	Low	High	High/ medium

Possible weaknesses with biometric tools

While it is not celebrated if a person's password is made public in a data breach, at least it can be easily change to a different password. However, if the scan of a person's fingerprint or iris (i.e., known as "biometric template data") is made known in the same manner, a person could be in a real dilemma or quagmire. After all, a person cannot replace his/her fingerprint or iris.

People's biometric template data are everlastingly and distinctively linked to them. The exposure of that data to hackers may seriously undermine user privacy and the security of a biometric system. Contemporary techniques offer effective security from breaches; however, advances in Artificial Intelligence is rendering these protections obsolete.

What is more, quite a few experts in the cybersecurity industry presume that adversarial types of Artificial Intelligence will be utilized to imitate individual users, and it is happening. That is, the Artificial Intelligence threat is based on the fact that security systems are based on data. Passwords are data. Biometrics are data. Photos and videos are data. Therefore, the threats to Artificial Intelligence is coming online that can generate replica data that passes as the bona fide item.

One of the most perplexing Artificial Intelligence technologies for security teams is a category of algorithms referred to as generative adversarial networks (GANs). GANs can imitate or simulate any distribution of data, including biometric data.

The word "adversarial" refers to a type that is constructed to deceive an evaluating neural net or another machine-learning model. With the implementation of machine learning in security applications increasing, this adversarial type has become very important. For example, documents with headers that contain either terminating tags, such as HTML, or document lengths, such as rich text formats (.rtf) or .doc file formats. Since these files may have arbitrary bytes appended to the end, it provides file space that could be implemented to produce these adversarial examples (https://securityintelligence.com/generative-adversarial-networks-and-cybersecurity-part-1/).

These adversarial threat actors may gain access to Artificial Intelligence tools that can enable them to defeat multiple forms of authentication—from passwords to biometric security systems, which includes facial recognition software that identify targets on networks and evade detection. Furthermore, threat actors will soon be able to basically shop on the dark web for the Artificial Intelligence tools they need to automate new kinds of attacks at unprecedented scale (https://securityintelligence.com/ai-may-soon-defeat-biometric-security-even-facial-recognition-software/).

Nonetheless, Artificial Intelligence and machine learning have displayed untold potential in cybersecurity. Likewise, Artificial Intelligence and biometrics can be combined to create a secure authentication mechanism. The combination of Artificial Intelligence and biometrics can help develop authentication systems that can protect devices against cyber-attacks and prevent fraudulent activities.

Artificial Intelligence and biometrics can work together, especially with the employment of keystroke dynamics, facial recognition, voice recognition, and gait detection. Keystroke dynamics and Artificial Intelligence can be combined to make keystroke dynamics more precise and a reliable typing pattern recognition technology. Artificial Intelligence systems can track information regarding how a person types and the time interval between two keys for the most habitually utilized keys to identify people. Moreover, Artificial Intelligence can learn from user profiles that are assembled over time as users continue typing.

Facial recognition can be more successful with the assistant of machine learning. Artificial Intelligence is trained from millions of images and uses 3D biometrics to effectively authenticate a person's face. In addition, Artificial Intelligence systems can execute predictive modeling to evaluate the effects of aging on the human face. To this end, Artificial Intelligence scrutinizes pictures of elderly people to recreate younger images of those people. Combining with the assistance of large volumes of available facial data, Artificial Intelligence and biometrics can together create more accurate authentication models.

The enactment of Artificial Intelligence in voice recognition can train the biometric systems utilizing millions of voice samples of different users. Artificial Intelligence and voice recognition can appraise an individual's biometric voice signature by studying their voice patterns such as speed, accent, tone, and pitch. Such biometrics can be fast; thereby, precisely authenticating people. Such Artificial Intelligence-powered voice recognition can be utilized in workplaces for authentication and attendance purposes.

The amalgamation of Artificial Intelligence and biometrics can improve the probability of correctly corroborating people's identities based on their physiological and behavioral traits. Nevertheless, their acceptance and putting into place are limited due to the lack of commercial applications. Hence, organizational leaders can introduce Artificial Intelligence-based biometrics for their workplaces and customers to offer user-friendly and secure authentication protocols (https://www.allerin.com/blog/biometrics-is-smart-but-ai-is-smarter-heres-why).

Privacy Issues and Security

Privacy has been an issue of concern with respect to biometrics for quite some time. The European Union's (EU) General Data Protection

Regulation (GDPR) highlights the right to explanation. Fundamentally, it mandates that users be able to demand the data behind the algorithmic decisions made for them. That is, this law provides individuals the right to view the data that companies hold on them, make corrections to it, and request that it be deleted and not sold to third parties.

In October 2012, the Federal Trade Commission issued a narrative, "Facing Facts: Best Practices for Common Uses of Facial Recognition Technologies," (https://www.ftc.gov/reports/facing-facts-best-practices-common-uses-facial-recognition-technologies) whereby The FTC staff made a number of specific recommendations regarding the implementation of biometrics tools. The recommendations embraced delivering clear notice about collection and use, giving users opting out rights; and obtaining express consent prior to using a consumer's image in a materially different manner than for which it was originally collected.

Although those issues were raised in 2012, they remain still unresolved. Nonetheless, a few individual states have legislated in this area. For example, Texas and Illinois have prevailing biometric privacy statutes that relate, in certain circumstances, to the collection, storage, and use of biometric information, including facial templates. In addition, the California Consumer Privacy Act of 2018 is similar to the EU's GDPR (https://iapp.org/news/a/gdpr-matchup-california-consumer-privacy-act/).

If a company decides to embrace biometrics technology, they have to determine whether to obtain prior affirmative consent (or post a physical notice) before collecting relevant data, as many mobile apps do before collecting geolocation data.

Data Security

Like any organization information, biometrics data can be subject to a security breach that results in sensitive data retrieved by unauthorized entities. In quite a few cases, the implications of a breach of biometric data may have an even larger influence than the breaches of financial and personal data that investors and creditors are confronted with presently. For example, a breach involving iris scans can have more grave consequences than passwords or payment card data, which can be changed or reissued, while an iris scan cannot (i.e., without surgery).

Organizations that collect and store biometric data must be prepared to contain it with appropriate levels of security, access restrictions, and safeguards. The cost of a breach could be significant, and organizations should estimate whether their cybersecurity insurance policies would cover their internal control systems.

Technology Failure

Similar to most new technologies, biometrics is not completely fool-proof. Moreover, although biometric safeguards are typically thought to be more difficult to hack, enterprising hackers are constantly attempting to "spoof" biometric readers to circumvent the technology. Therefore, in considering whether to implement these technologies, an organization ought to consider the implications if an application erroneously misidentifies an individual (i.e., Type 2 error), whether through malfeasance or simple technology failure. A Type 2 error (false accept rate or false match rate) transpires when a biometric system incorrectly authenticates the identity of an individual and grants them access to what the system is safeguarding (Rodgers, 2012). Whereas a Type 1 error refers to non-acceptance of biometric feature, which ought to be accepted.

Economics

State of the arts biometrics systems can be expensive. Depending on the system and its employment, it may involve a great deal of information technology investment (i.e., in such add-ons such as software and hardware, legacy system upgrades, maintenance, and consulting) as well as employee training and user education. Nonetheless, there are many expected benefits. Considering the use of biometrics, high costs should be regarded in the context of the expected benefits.

User Perception

While many believe that the proficiencies of biometrics can be very beneficial, many are also ill at ease regarding the application of the technology. Therefore, organizations have sought to begin to evaluate the use of valuable (but potentially invasive) technologies like facial, iris, or voice recognition taking into consideration the customers, suppliers and others' reactions.

With the high level of awareness about privacy and the requirement for data security engendered by internal control breaches, will customers worry about their biometric data being collected, stored, and processed?

On the other hand, many are impressed with the technological dimensions of the utilization of biometrics. Hence, to the extent an organization pursues a technological upgrade, the current image with customers and other stakeholders regarding the use of biometrics may help.

Biometrics for Warehouse Security

Biometrics can be applied in any type of environment and are used for the benefit of the public. A good example where using biometrics will benefit both customers and vendors would be a warehouse. A warehouse is used to store goods that can widely range in prices. The use of biometrics in a warehouse can help in protecting the valuable goods. All warehouses have some type of security such as security guards who patrol the property where the warehouse is located. Some other type of security measures are cctv to monitor the activities outside and inside the warehouse and some type of fencing to help protect the valuables. Other common types of security a warehouse may utilize is an alarm system. The alarm system can be a pin pad where a code is entered, or a card key will be used that will be paired up with a card key reader to activate or deactivate the alarm system. The primary goal of the vendors that hold their merchandise in a warehouse is to keep unauthorized personnel out. Unauthorized personnel will be either lost or attempting to take merchandise that is of no use to them. In order for vendors to safeguard their goods they should implement good security features.

The use of biometrics combined with security systems already in place can help decrease the number of individuals who can get access to the warehouse without permission. Some common types of biometrics that can be used for security measures include fingerprint recognition, iris scan, gait recognition, face recognition and even voice recognition. The type of biometric that can be used for security in a warehouse will depend upon the amount of value of the goods stored in the warehouse. Also, the value of goods will determine whether the organization desires to emphasize more control on Type 1 or Type 2 errors. If the value of the goods stored in the warehouse is high then it is worth spending a high amount on the security. Vice versa, if the value of the goods stored at the warehouse is low then it might not be worth spending a lot of money. For example, if the warehouse is holding the stock of a grocery store or a retailer and the per item price is very low then it might *not* be worth spending a great deal of resources on biometric tools to increase the security. On the other hand, if the warehouse is storing inventory that is very expensive such as luxury automobiles, special machinery parts or something as valuable as "confidential information" then it is definitely worth spending money on biometric tools to increase the security. The most affordable biometrics that exists at the present time is finger print recognition. Finger print recognition can be used to identify and verify someone's identity. Finger print recognition is very affordable since it is a small portable machine that can be easily moved around and does not require a large system. The process requires everyone's finger prints to be stored into a database, once they are in the database then when someone scans his/her finger, it will either match it or not match it to a finger print already in the database. This biometric does not relate to deep learning, such as trying to find more algorithms or doing more research beyond what the database contains. This biometric can be easily fooled by taking someone's finger print and pretending to be someone they're not. The other biometrics systems are more complicated and very expensive. A highly effective biometric is the iris scan. In this biometric tool the person looks into a camera that will scan his iris and compare it to the scans in the database. It is not easy to fool this type of biometric and therefore is more effective when it comes to identifying a person.

This biometric will be rule-based because there is no deep learning procedure other than within its own system, and there is no deep learning outside of the database. The information is already recorded in the system; therefore, it will either match or not match the finger print or iris scan. Once a person scans his/her finger print or iris, the matching mechanism will approve it according to his/her records or not approve it, there is no in-between. It will be a complete match or be denied entirely. The utilitarian based principle is another strategy to employ in that it is for the best interest of the stakeholders (e.g., shareholders). Utilitarianism is described as the best interest for a particular group of stakeholders; therefore, maximizing the good for a particular group. The use of biometrics for security purpose is set up for the protection of the goods of the company, which might conflict with existing rules or laws.

To sum up, the use of biometrics for the security for expensive items in a warehouse would lean more towards a Type 1 error rather than a Type 2 error. In addition, a determination for the two types of errors should be calculated in order to determine the types of biometrics tools to be installed in a warehouse. In this case, a Type 1 error could reject the entrance of an un-authorized personnel due to more strict measures. Any relaxation of security may promote more of type 2 error (i.e., accepting the entrance to un-authorized personnel).

A decision-tree for the biometrics used in warehouse security illustrates how existing security works as a two-factor authentication (Figure 4.17). Regular security will stay in place to help correctly identify authorized personnel and the biometric will be used to verify the identification of the personnel. The decision tree will start with a simple task to see if the person trying to enter the warehouse is authorized or un-authorized, this will branch out into two simple answers, either yes or no. If this person is identified as un-authorized personnel, then the door or gate to enter the warehouse will not open. The other option is yes, if this person was identified as authorized there needs to be further investigation, how was

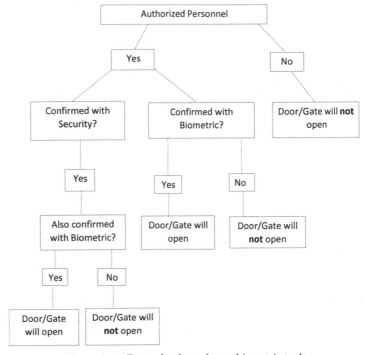

Figure 4.17. Example of warehouse biometric tools

this person identified. If s/he identification occurred through regular security measures, it needs to be verified with a biometric tool, if it is not verified with a biometric tool the gate or door will not open. If it is verified with a biometric tool and the result is a match, the door or gate will open, allowing access to the person. If the biometric does not match any finger print (or iris scan) in the system, the personnel will be labeled as un-authorized personnel even if the security proved otherwise; therefore, the door will not open again. The end result is that it will always need to be a match with the biometric in order for the person to gain access into the warehouse. Using biometric tool as a second verification of identity will provide the best result to keep un-authorized personnel outside of the warehouse. The use of the biometric can begin with any type of warehouse facility such as gait, iris or finger print recognition. Warehouses can decide if the biometrics is affordable, and as time passes on, they will decide whether there needs to be an upgrade into something more accurate.

As technology is advancing the use of biometrics will also advance, biometrics can be applied at anywhere. Some may be more complicated than others; however, they will definitely help reduce the amount of fraud that happens in a workplace. These biometrics can take different forms by applying information available and coming to a decision weighting many factors. Artificial Intelligence coupled with biometrics will continue to advance and will become more sophisticated as technology advances and more discoveries are made.

In addition to utilizing biometric technologies to improve workflow in warehouses, organizations are looking at its capability to enhance security as an access control system as well. Nonetheless, some organizations have integrated biometric technology across many more systems such as time and attendance tracking for employees.

Conclusion

With regards to biometrics, organizations must rise to the challenge or be disadvantaged relative to traditional and non-traditional competitors. Artificial Intelligence, automation, and analytics employing biometrics are central to the success of organizations, and will pervade critical business and non-business areas, including data, business processes, the workforce, and risk and reputation.

Undoubtedly, biometrics will be part of the commercial and non-commercial environment in the future, and the timing of biometric investments in this area is key. Interested organizations must remain aware of the emerging legal landscape, and careful contracting is essential to protect investments in the application. Of course, as a matter of both internal practice and third-party contractual relationships, emphasis should be placed on data security. Finally, it is important for organizations to understand the risks and retain flexibility to maintain compliance as the legal and commercial environment evolves.

As individuals are becoming more and more connected online and offline, there is no question that our digital consumption will tell more and more about ourselves. This may allow better short-term projections of our habits and interests. Nevertheless, the challenge of forecasting future needs and behavior is much more difficult. Therefore, the drive for big data in organizations will increase over time.

Moreover, much has been made about the fact that Artificial Intelligence's reliance on big data is already impacting privacy in a major way. Without proper regulations and self-imposed limitations, critics argue that the situation may become regretfully unhealthier.

Biometric identifiers along with big data are related to intrinsic human characteristics. They basically can be grouped into two categories: physical identifiers and behavioral identifiers. Physical identifiers are, for the most part, innate, immutable and device independent. Whereas, behavioral identifiers relate to the measure of uniquely identifying and measurable patterns in human activities.

Recent advances in the field of Artificial Intelligence research are gaining widespread attention from the world over because of the impact that they can have on our lives. Artificial Intelligence is more and more combining with biometric tools, especially as it pertains to physical, behavioral and multimodal biometrics. In an Artificial Intelligence biometric system, the Artificial Intelligence can account for and learn subtle changes (i.e., via machine learning or deep learning) in an individual to be authenticated.

Moreover, biometric control systems that uses Artificial Intelligence can also be used for continuous biometric monitoring. For example, in a chiefly secure location where a user regularly labors on documents, an artificial intelligence may constantly monitor the user's typing behaviors to verify that no data tampering is happening and that the individual editing a document is, consistently, who they say they are.

Artificial Intelligence systems can be used in tandem with other, more traditional forms of biometric authentication such as iris and retinal scans. Advances in Artificial Intelligence can also benefit these mature yet effective methods of biometric authentication. For example, machine vision has become notably accurate and can assist some methods that have been hitherto considered intrusive more desirable to users, faster, and more practical.

To sum up, biometric systems assume and necessitate an intimate relationship between individuals and technologies that collect and record the biological and behavioral characteristics of their bodies or movements. Hence, it is incumbent upon those who conceive, design, and deploy biometric systems to ponder the cultural, social and legal contexts of these systems. Not attending to these considerations and failing to consider social impacts weakens their efficacy and can bring grave unintended consequences.

References

Breur, T. 2016. Statistical Power Analysis and the contemporary "crisis" in social sciences. Journal of Marketing Analytics, 4(2-3): 61–65.

Goes, P.B. 2014. Design science research in top information systems journals. MIS Quarterly: Management Information Systems, 38(1): iii–viii.

Laney, D. 2001. 3D data management: Controlling data volume, velocity and variety. META Group Research Note: 6(70).

Rodgers, W. 1993. The effects of informational complexity on lending officers' decision processes. Accounting Enquiries, 2: 363–404.

Rodgers, W. 2006. Process Thinking: Six Pathways to Successful Decision Making. NY: iUniverse, Inc.

Rodgers, W. 2010. E-commerce and Biometric Issues Addressed in a Throughput Model. Hauppauge, NY: Nova Publication.

Rodgers, W. 2012. Biometric and Auditing Issues Addressed in a Throughput Model. Charlotte, NC: Information Age Publishing Inc.

Rodgers, W. 2016. Knowledge Creation: Going Beyond Published Financial Information. Hauppauge, NY: Nova Publication.

Strong, C. 2015. Humanizing Big Data: Marketing at the Meeting of Data, Social Science and Consumer Insight. London: Kogan Page.

Tversky, A. and Kahneman, D. 1974. Judgments under uncertainty: Heuristics and biases. Science, 185: 1124–1131.

5

Ethical Issues Addressed in Artificial Intelligence

"What does the notion of ethics mean for a machine that does not care whether it or those around it continue to exist, that cannot feel, that cannot suffer, that does not know what fundamental rights are?"
— Vijay Saraswat, Chief Scientist for IBM Compliance Solutions

Due to the tremendous complexity of the world around us and the enormous amount of information in the environment, it is necessary sometimes to rely on some psychological shortcuts or heuristics that allow us to act expeditiously based on the circumstances. Moreover, the escalating use of algorithms in decision-making also brings to light important issues about governance, accountability and ethics.

Why the increase in popularity of Artificial Intelligence ethics and Artificial Intelligence governance? Artificial Intelligence technology is displaying disruption in business and life in general. In other words, Artificial Intelligence and machine learning are recognized as disruptive forces that will likely change the human experience, and in particular, the nature of human work.

Since Artificial Intelligence is recognized as a colossally disruptive force, people are interested in guaranteeing that its impact on society and individuals is appropriate, respectable and beneficial. Certainly, much of the origin of "Artificial Intelligence Good" initiatives originate from a desire to do good (Rodgers, 2009).

Nevertheless, obtaining power has impacted upon Artificial Intelligence ethics. Individuals, organizations, and nations are now realizing just how serious their disadvantage will be without Artificial Intelligence innovation. For these groups, securing one's interests in the future suggests a corridor other than innovation, and regulation is the next best thing.

The most essential principle of power and Artificial Intelligence is data dominance. That is, whoever controls the most valuable data within a space or sector will be able to make a superior product or solve an advance problem. Whosoever solves the problem best will be dominant in business, secure higher revenues, and whoever wins customers obtain more data.

For example, Google is likely to obtain more general search queries, and therefore, people will not likely use any search engine other than Google. Hence, Google obtains more searches (data) to train with, and gets an even better search product, which could lead to a search monopoly.

Further, Amazon generates more general eCommerce purchases than any company; therefore, people will not likely use any online store other than Amazon. Hence, Amazon obtains more purchases and customers (data) to train with and becomes an even better eCommerce product. Thereby resulting in eCommerce monopoly. On the other hand, Artificial Intelligence may alter other companies such as Facebook, Netflix, Uber, etc., and they become less reliant on data collection, and data dominance, which may cause them to eventually be eclipsed by some other power dynamics.

Society generally aims to be good and hope that others will follow suit. Nevertheless, individuals and organizations often times do what behoves their own interests. Perhaps a solution to the problem may be to build ethical governance mechanisms that take this inherently amoral egoism into account. Therefore, this chapter will center on an ethical model that addresses six dominant algorithmic pathways in Artificial Intelligence-oriented decision-making. This chapter reviews the following areas: (1) ethics, Artificial Intelligence tools and biases, (2) ethical issues depicted in the Throughput Model algorithmic pathways, (3) Artificial Intelligence ethical regulations, (4) root case analysis for Artificial Intelligence biometrics systems, (5) data classification system for Artificial Intelligence ethical algorithmic pathways, (6) hospitals' implementation of biometrics, and (7) caveats pertaining to privacy rights.

Ethics, Artificial Intelligence Tools and Cognitive Biases

From machine learning and deep learning, algorithmic stock trading to social security compensations and risk management for loan approvals has demonstrably increased the use of Artificial Intelligence. The concept of the algorithmic accountability suggests that organizations should be responsible for the results of the programmatic decisions. Ethical decisions of Artificial Intelligence, suggest a need to secure "ethical" training datasets, and well-designed boundaries to govern "ethically" Artificial Intelligence decisions.

Algorithms are basically a set of instructions that can be understood by a computer to solve a problem or complete a task. Algorithms are not inherently biased, algorithmic decision choices depend on a numerous factor. Additional factors include how the software is employed, and the quality and representativeness of the underlying data. Data transparency, review and remediation should be taken into account during the course of algorithmic engineering processes.

A heuristic is a rule, strategy or similar cognitive shortcut that a person implements in order to derive a solution to a problem. A heuristic that works all of the time is known as a type of algorithm. A systematic error that results from the use of a heuristic is called a cognitive bias (Rodgers, 2006).

Big data is typically a representation of financial and non-financial data. Further, some of the non-financial data represents survey data. Pertaining to the use of big data through Artificial Intelligence applications, the typical two major types of biases are: (1) *selection bias* and (2) *response bias*. *Selection biases* that may arise include non-representative sample, non-response bias and voluntary bias. *Response biases* refer to the various conditions and biases that can influence survey responses, which can be used as part of big data. The bias can be intentional or accidental, but with biased responses, survey data becomes less meaningful as it is inaccurate.

The common underlying factor in cognitive biases is predispositions towards an idea, values, attitudes or beliefs. There are many cognitive biases that may impact on individual or organizational decision-making (Rodgers, 2006). Emotions such as anxiety, fear or anger could easily cloud a person's judgement. Human thinking is prone to the use of cognitive heuristics, a shortcut that may lead to biases and faulty decisions.

Susceptibility to bias in Artificial Intelligence is inspired through the assignment of weight on the parameters and nodes of a neural network, a computer system fashioned on the human brain. The weight may inadvertently bias the machine learning algorithm from initiation by means of data input, through supervised training, and by involvement through manual adjustments. The deficiency or inclusion of indicators and the inherent cognitive biases of an individual computer programmer may instigate machine learning bias (Rosso, 2015).

For example, in a baseball learning algorithm seeking players who are most likely to succeed, the training dataset may be fed with data of 200 profiles from the top-performing players in the baseball organization. The algorithm then seeks out patterns and correlations, which promote to their predictive power when scrutinizing the likelihood of success in a new baseball player, based on their profile. Handing decision-making over to machine learning algorithms has many benefits for the individuals in question, including saving time, money, and effort. Nonetheless, when it comes to the ethics and responsibility of the decision choice, the lines may become blurred. Since we are not able to absorb correctly why a machine may have made the decision choice that it did, we are not always able to detect and escape biases when it occurs. It is situations such as the above that a proper understanding of ethical positions is necessary in the decision-making engineering of Artificial Intelligence software.

The areas of cognitive bias vulnerability for machine learning encompasses (Rosso, 2015):

(a) Data structure, collection and sources;
(b) Data set size;
(c) Stage of objectivity in the data;
(d) Weight assignments to data points;
(e) The deficiencies or inclusion of indicators; and
(f) The intrinsic cognitive biases of the human programmer.

Echoing larger cognitive biases and heuristics dynamics in the global community, this chapter suggests that Artificial Intelligence and the proliferation of ethical codes and principles challenge "traditional" aspects of professions and professional norms should provide opening up windows for new movements towards professionalization. Therefore, this chapter utilizes an ethical decision-making process, which is based upon the Throughput Model (Rodgers, 2009). This process includes six dominant algorithmic ethical pathways to a decision choice. Discussed later in this chapter, the six dominant ethical considerations that are tied to algorithmic pathways are divided into primary and secondary positions. The three primary ethical positions are: preference-based (e.g., ethical egoism), deontology or rule-based (rights, laws, guideline, and procedures), and utilitarian or principal-based (values, attitudes and beliefs). The three secondary level positions are relativist, virtue ethics, and ethics of care (Rodgers and Al Fayi, 2019; Rodgers and Gago, 2001, 2003, 2004; Rodgers et al., 2009; Rodgers et al., 2014).

The aforementioned ethical considerations are important to understand because bias can exist in people as well as Artificial Intelligence platforms. Bias can be defined broadly as a deviation from some rational decision choice or norm, and can be statistical, legal, moral, or functional. Bias exists in our everyday lives as well as on a societal scale. Now and again, one perpetuates the other.

Why use the Throughput Model Algorithmic Pathways for Ethical Considerations?

Envision, in the not to distant future, an insurance company implementing a machine learning algorithm to endorse car insurance applications for approval. Unearthing an answer may not be as trouble-free. If the machine learning algorithm is grounded on a complicated neural network, or a genetic algorithm fashioned by directed evolution, then it may prove quite impossible to understand why, or even how, the algorithm is evaluating applicants based on their geographical location. On the contrary, a machine learner based on regressions, decision trees or Bayesian networks is much more transparent to programmer inspection (Hastie et al., 2001). Therefore, Throughput Modelling enables an inspector or auditor to discover that the Artificial Intelligence algorithm uses a particular pathway to address other information of applicants as opposed to who were born or previously resided in predominantly poverty-stricken areas.

When Artificial Intelligence algorithms take on cognitive work with social dimensions-cognitive tasks heretofore accomplished by people, the Artificial Intelligence algorithm inherits the social requirements. The Throughput Model six dominant pathways allows for the development of Artificial Intelligence algorithms that are not just powerful and scalable, but also *transparent to inspection*, which is one of many economically and socially essential properties.

Transparency is not the only appropriate characteristic of Artificial Intelligence. It is also of great consequence that Artificial Intelligence algorithms taking over social functions be *predictable to those they govern*. Further, the task of decision-making is not necessarily to optimize society, but to stipulate a predictable environment within which citizens can optimize their own lives. Therefore, the Throughput Model provides six dominant algorithms in order to accomplish the aforementioned problem.

The algorithmic ethical decision-making processes can be represented in an organized manner. In order to study the ethical considerations of algorithms, it is important to break up all the paths marked with arrows in Figure 5.1 into sets of individual algorithmic ethical pathways. These algorithms can then be independently analyzed for their contributing properties to individuals' decision processes (Rodgers, 1997). Further, it is common for individuals or organizations to differ in their moral philosophical values. For example, even though two people agree on the ethical principles that determine ethical behavior, it is unlikely that they will agree on the relative importance of each principle. These differences are highlighted in Figure 5.1, depicting several algorithmic pathways towards making a decision.

Based on Figure 5.1, we can establish six algorithmic ethical pathways (Rodgers, 2009):

(1) **P** → **D** *ethical egoism (preference-based) position*
(2) **P** → **J** → **D** *deontology (rule-based) position*
(3) **I** → **J** → **D** *utilitarian (principle-based) position*
(4) **I** → **P** → **D** *relativist position*

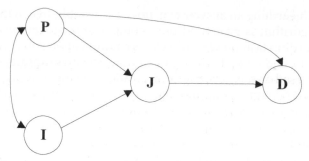

Figure 5.1. Throughput Modeling: Ethical process thinking model (Rodgers, 2009). *Where,*
P = perception, I = information, J = judgment and D = decision choice

(5) **P → I → J → D** *virtue ethics position*
(6) **I → P → J → D** *ethics of care (stakeholder-based) position*

This *Ethical Process Thinking Model* (Rodgers, 2009; Rodgers and Al Fayi, 2019) illustrates: (1) only 2–4 major concepts that are instrumental in reaching a decision choice; (2) the order of a particular pathway (and its strength) will greatly influence the outcome of a decision; and (3) each decision-making pathway relates to a particular ethical position. There are many philosophies, and are complex in nature. We shall discuss six prominent algorithmic approaches depicted in the model's pathways. The six ethical positions discussed below are ethical egoism (preference-based), deontology (rule-based), utilitarianism (principle-based), relativist, virtue ethics, and ethics of care (stakeholder-based).

(1) **P → D** indicates *ethical egoism position*, which emphasizes that people are always motivated to act in their perceived self-interest. Utility theory or Game theory supports this position. That is, Game theory is the study of the means that strategic interactions among rational players produce outcomes with respect to the preferences (or utilities) of those players, none of which might have been intended by any of them.

(2) **P → J → D** represents the *deontology position* that stresses the rights of individuals and on the judgments associated with a certain decision process rather than on its choices. That is, this decision-making position indicates that one's perception (P) is oriented by conditioning judgment (J) of the rules and laws before arriving at a decision (D). That is, the decision choice is brought on by a judgment based on a perception of a circumstance. Moreover, deontology calls attention to individual rights pertaining to for example, freedom of speech. Even so, generalized rules of ethics may be unhelpful in counselling students. In other words, students who know the rules of law, but

have no developed personal beliefs about their own duty to obey them, have not been fully educated in ethical considerations.

(3) $I \rightarrow J \rightarrow D$ echoes the *utilitarian position*, which is concerned with consequences, as well as the greatest good for the greatest number of people. Utilitarianism is based on collective "economic egoism". The judgment is based on information and the information conditions the decision. This position promotes egalitarianism in the merged impacts of economic growth and civil development of social citizenship principles. That is, the civil development consists of a person's liberty, freedom of speech, thought and faith, the right to own property and to determine valid contracts, and the right to justice. Utilitarianism is an extension of ethical egoism in that it is committed to the maximization of the good and the minimization of harm and evil to a society.

(4) $I \rightarrow P \rightarrow D$ is symbolic of the *relativist position* that assumes that decision-makers use themselves or the people around them as their foundation for defining ethical standards. A conflict of values and interests, and tensions pertaining to what is and what some groups believe can discourage accommodations with other interested parties. Ethical relativism is the position that maintains that morality is relative to the norms of one's culture or the situation. An action that is right or wrong lies upon the moral norms of the society in which it is exercised. The same action may be morally right in one society; however, it may be morally wrong in another. For the ethical relativist, there are no universal moral standards since there are standards that can be universally appropriate to all individuals at all times.

(5) $P \rightarrow I \rightarrow J \rightarrow D$ highlights the *virtue ethics position*, whereby the nurturing of virtuous traits of character is considered as morality's primary function. Many philosophers are quite explicit when they state that a wicked person is in charge of his/her character. That is, not because he/she could now transform it but because he/she could have and should have acted differently early on and established very dissimilar habits and states of character. In the virtue ethics position, a circumstance is perceived (**P**). A conscious look for information (**I**) is initiated. Based on the information a judgment (**J**) is made, which will support a decision (**D**).

(6) $I \rightarrow P \rightarrow J \rightarrow D$ denotes the *ethics of care position* (stakeholders' perspective) that centers on a willingness to listen to distinct and previously unacknowledged viewpoints. In other words, an organization must build solidarity among its employees, suppliers, customers, shareholders, and the community. The ethics of care position acknowledges the moral priority of caring for the distinctive needs of others for whom we are responsible. This stakeholder position focuses on responsiveness to need, empathetic understanding, and

the interrelatedness of people, rather than on individual rationality or universal moral rules. It places prominence on relations between people rather than the preferences or dispositions of individuals; it is thoughtful relations that are thought to have primary value.

Ethical Regulation Pertaining to Artificial Intelligence

In addition to the above, another area where this chapter contributes to the current debate about professional and ethical norms around Artificial Intelligence is by mapping and discussing the six dominant ethical positions and algorithmic pathways of governance effects on organizations. Specifically, we examine how six ethical decision-making algorithmic pathways interface with different types of accountability mechanisms. Moreover, examples are provided in terms of some of the ethical pathway positions implementation into practice. Further, this chapter highlights the interfaces between ethical considerations and other accountability mechanisms, including the court of public opinion and the legal system.

Artificial Intelligence producing algorithms, machine learning and deep learning are questioned since they are deemed to be operating from the public view. Technologists, politicians and academics have all exclaimed for more transparency centered on the systems implemented by dominant tech companies. In other words, oversight and transparency is called for when algorithms influence individuals' rights, public values or public decision-making. In addition, quite a few social media platforms are implementing a variety of different automation forms in distinct manners that limit people's ability to express themselves online.

One of the more essential sections of the European Union's groundbreaking General Data Protection Regulation (GDPR) spotlights on the right to explanation. Fundamentally, it mandates that users be able to demand the data behind the algorithmic decisions made for them (https://iapp.org/resources/article/the-eu-general-data-protection-regulation/). This includes but is not limited to recommendation systems, credit and insurance risk systems, advertising programs, and social networks. Therefore, the GDPR undertakes "intentional concealment" by organizations. Nevertheless, it does not remedy the technical challenges accompanied with transparency in modern algorithms.

Therefore, an ethical decision-making model may provide valuable input by analyzing various inputs used by a decision-making algorithm. Moreover, ethics is a foundational building block of human socio-economic dynamics. This relates to the primary challenges that people envision that algorithms are more than codes. That is, transparency is not the same as accountability. Further, the forces driving algorithmic systems are very dynamic, complex and probabilistic. This makes a diminutive impression to discuss the algorithm as if it were a solitary fragment of code that is not continually transforming, which means different things to different individuals.

Many of the difficulties that we currently view with an algorithm's fairness are problems with how we define ethics, and what kind of equitable outcomes we are attempting to reach. While it is challenging to audit algorithms, it is not unattainable. More demanding questions arise regarding who should define what is fair. Defining a fair process cannot and should not be left without considering a decision-making model along with an ethical component. To properly regulate technology, which influences lives, we ought to regulate the decision choices that technologies are putting into their algorithms. Although, this may not be a simplistic issue. Therefore, of primary importance is maintaining that algorithms are accountable through human intervention, auditing, and strong government regulation. This is necessary to ensure that we have equity and faith in automated decision-making processes advancing forward.

Auditing is an essential function for organizations, but much of it is routine. It lends itself to the use of technology to analyze big data and decide which areas of the audit to focus upon and how best to gather the data needed to ensure the audit meets professional and ethics standards. Accounting firms are experimenting with Artificial Intelligence, in which machines go beyond doing rote tasks and inform basic decision-making.

There are several implications of Artificial Intelligence for the corporate governance system including internal auditing, internal control over financial reporting, and the role of the external auditors. Moreover, unintended consequences may exist with respect to the potential for fraud in Artificial Intelligence systems and developing efficient and effective Artificial Intelligence systems.

According to an article by Steven Mintz article (https://www. ethicssage.com/2018/05/artificial-intelligence-ai-changing-the-way-audits-are-conducted.html) on how Artificial Intelligence is changing the way audits are conducted suggests that the following elements must be addressed in order to improve ethics in Artificial Intelligence:

1. How can an organization inaugurate accountability and oversight through corporate governance systems in an Artificial Intelligence environment?

2. How does the utilization of an Artificial Intelligence system stimulate the ethical role and responsibilities of the CFO? This can be linked to the requirement to verify financial statements under SOX Section 302.

3. Can Artificial Intelligence systems be implemented to enhance the agenda of management that may encompass occupational and/or

fraudulent financial statements. If yes, how can the integrity of the financial statements be protected?

4. What competencies will be required in order to facilitate internal auditors to promote Artificial Intelligence-related advisory and assurance services?

5. How can Artificial Intelligence systems be utilized to aid specific audit routines such as accounts payable?

6. How can Artificial Intelligence change the configuration of the audit teams in accounting firms?

7. What is the role and responsibilities of the system of internal controls over financial reporting in an Artificial Intelligence environment? This can be associated to management's obligation to evaluate whether internal controls in an Artificial Intelligence system are functioning as intended and the external auditor's role in reviewing management's report and coming to an independent assessment, as ordained under Section 404 of SOX.

8. What is the role and responsibilities of the audit committee in an Artificial Intelligence environment?

9. What are the probable threats to objectivity and integrity in an Artificial Intelligence environment and the defenses to moderate those threats and enhance the external audit function?

10. What role might the fraud triangle take part in the prevention and detection of fraud in an Artificial Intelligence environment?

11. What is the role of machines that go beyond doing rote tasks to make decisions that humans once made?

12. What are some of the societal/ethical issues of using Artificial Intelligence systems.

    a) Will robots take over some of the responsibilities of accountants?

    b) What specialized training is compulsory in the audit environment?

    c) What are the inferences for data privacy and the ethical use of personal data?

    d) What are the associations of utilizing Artificial Intelligence systems to audit with regard to the requisite to exercise a degree of professional skepticism that is indispensable in meeting the objectivity and due care standard in professional codes of ethics?

Root Case Analysis for Artificial Intelligence Biometrics Systems

Organizations are increasingly implementing machine learning and deep learning models to make decision choices. Examples include job allocation, loans, university admissions, etc., that directly or indirectly influence an individual's life. Algorithms are also implemented to recommend a movie to watch, a person to date, or an apartment to rent as well as to purchase a house. This type of technology advances has affected every industry in modern society and one of them is the healthcare industry. Technology has had a huge impact in the way the healthcare industry has been modernized and organized by changing the way records are being stored. Earlier, physicians had to keep track of numerous files for every patient that was examined, and it was very easy to misplace and mix the files with other patient's records. Electronic health records have had significant savings for the industry costs, and it has assisted in improving the patient's health and safety by implementing biometric measures in the workplace.

Biometric authentication is the identification of a person's unique physical or behavioral characteristics. Technology advances has made it easier to identify a person by implementing biometric machines or scanners that in seconds can match a persons' physical or behavioral characteristics with an individual based on a set of databases that have been stored. Since these databases contain private information it is very important that every health insurance company within the industry establishes good internal controls to prevent any unusual activity that could affect employees, patients and stakeholders within the industry. Biometrics and internal controls relate to each other by securing and minimizing risks.

Ineffective internal controls within the healthcare industry does not only affect patients but also nurses, doctors, administrators, visitors, contractors and all the staff. These risks can be mitigated with biometric measures, but the use of biometrics has many different points of views in terms of how ethical the people within the industry are. There are four different types of ethical perspectives that will be analyzed in this case which are deontology, utilitarianism, relativism and ethics of care.

Social norms

Social norms are the expectation of human behaviors. Society is capable of deciding what rules to follow and how to govern the behavior of the members of a society. As mentioned earlier, in the healthcare industry there are many regulations that protect the public from unethical behaviors that could lead to fraud and affect a lot of people involved in the industry. Implementing biometric methods by healthcare providers is essential to have an effective control over the industry. For example, requiring staff to present their credentials while entering a healthcare provider facility is a way of implementing biometrics. In some cases, people may think that this requirement is unnecessary and as simple as it seems to carry a badge, they opt not to do it. In this industry, healthcare providers have access to a person's financial status and that can give them an incentive to give wrong diagnosis in order to keep the patient for a long period of time, this action is unethical. Healthcare providers can also decide to not cover a certain treatment, but they need to consider how this decision affects society. Healthcare providers have to measure the benefits and costs of covering a treatment.

From a biometric point of view, healthcare providers invest in biometric tools to protect their patients and staff. Such investments can be very expensive depending on the type of biometric equipment use but the decision to acquire the necessary equipment is determined on how it will benefit them. Implementing facial recognition and fingerprints to check that the physicians and staff attending to the patients are the individuals they claim to be, will benefit the company because it limits the risk that another person without the necessary knowledge will be treating a patient and giving the wrong information. Another type of biometric that healthcare providers use is DNA recognition. This method can be performed easily and obtained from different biological sources such as body fluids, nail, and hair. DNA recognition is painless, convenient and simple to perform making it beneficial to the healthcare provider and the person being tested.

Data Classification System for Artificial Intelligence Ethical Algorithmic Pathways

There are five confidentiality levels to classify data in an Artificial Intelligence ethical algorithm system (Rodgers, 2019). Level 1 represents data that will not cause harm to people inside or outside of an organization. Level 2 depicts information that the reporting entity has decided not to disclose; but, would cause any harm if it were made public. Level 3 is somewhat more serious in that if information is leaked from the entity then there could be some material harm to the organization or individuals. This level represents no real health harm; however, it may result in stolen passwords, identification or important documents. Level 4 pertains to if some information was released than individuals or the entity would suffer harm. This indicates that harm relates to contact with the person or entity causing physical harm. Finally, level 5 means severe harm to any individual or entity if the information was disclosed. These levels will be presented in the use of deontology, utilitarian, relativism or ethics of care systems described in the following sections.

Deontology

Deontology deals with the laws, procedures, guidelines and regulations that have to be considered in order to implement an Artificial Intelligence driven biometrics system in the workplace (Figure 5.2). In this scenario, the healthcare industry is in demand of biometrics for staff authentication and patient identification. Facial recognition and fingerprints can be required as access identification. These biometrics can assist to keep confidentiality by restricting the sharing of credentials, passwords as well as badge IDs.

Deontology consists of the following algorithmic pathway: perception, judgement, and decision. This ethical position is influenced by following certain rules or regulations. The decision is reached based on judgement. Deontology relates to the healthcare industry in many ways. There are many regulatory agencies within the healthcare industry such as the US Department of Health and Human Services, Centers for Medicare & Medicaid Services, Health Insurance Portability and Accountability Act and the Health and Safety Authority, which are responsible for maintaining health and safety in the workplace. These agencies help the public by monitoring the health status within a community, they diagnose and investigate health hazards and inform others of health issues. In this industry, everyone including patients have the right to privacy and healthcare providers should be able to provide discretion and never use sensitive information for their interest. The Health Information Technology for Economic and Clinical Health Act (https://www.hhs.gov/hipaa/for-professionals/special-topics/hitech-act-enforcement-interim-final-rule/index.html) is one of the regulations that demand IT professionals to be present in the industry. This act notifies patients of any unsecured data breaches that might compromise their privacy. If a person is in a situation where they have to make a decision where they are compromising the privacy of others, thinking about these regulations could help them make a better ethical decision by considering the consequences of a violations. Under the Health Insurance Portability and Accountability Act there are mandatory penalties for willful neglect; these penalties can result in a fine of $250,000. To prevent violation healthcare providers should have biometric tools to assist in the identification of the personnel with access to sensitive information. One biometric method that can be utilized is facial recognition, which has become very popular in the market because of its success in matching physical characteristics to pictures already stored in a database. In order to identify that the right person is accessing the database fingerprints, finger geometry or voice recognition can be used because they are considered unique for each person.

What happens if the biometric authentication system makes an error in properly identifying a person? There are two main types of performance metrics: *Type 1* errors (false reject rate or false non-match rate) and *Type 2* errors (false accept rate or false match rate). Type 1 errors occur when the biometric system rejects the identity of an individual that has authority to access what is being protected. Type 2 errors are opposite in that a

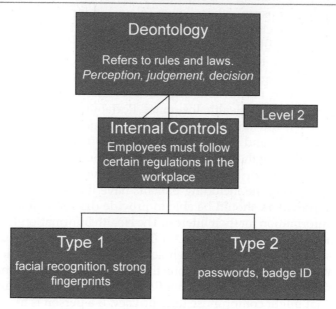

Rules

Recent researchers have estimate that the use of biometrics in the healthcare market will increase the demand for biometrics to be used for healthcare fraud prevention and cost containment in the US.

Deontology

Refers to rules and laws.
Perception, judgement, decision

Internal Controls

Employees must follow certain regulations in the workplace

Level 2

Type 1

facial recognition, strong fingerprints

Type 2

passwords, badge ID

Figure 5.2. Deontology algorithmic system

biometric system *incorrectly* authenticates the identity of an individual and grants them access to what the system is protecting (see Figure 5.2). In the decision tree example in the figure, the cash inventory held within the safe is very valuable. It would be preferable to have a Type 2 error where an authorized person is denied access to the cash. A Type 1 error could be detrimental to the financial stability and reputation of an organization in a situation where an unauthorized individual access the safe and takes all the money. Type 1 errors may be preferable in situations where high value items are not at stake and occasional unauthorized access is preferable to inefficient processing. An example of this would be access to a movie theater. A movie ticket is not an expensive item and the movie theater will not go out of business if an unauthorized individual sneak into a movie without paying. However, the movie theater may not be able to sustain business if it implements complicated security systems for customers to access the theater rooms and it becomes inconvenient for customers to watch their movies.

Utilitarianism

Utilitarianism is determined by the following pathway: Information, judgement and decision choice. Utilitarianism is somehow related to the ethical egoism position because they are both concerned with the consequences of their decisions. Utilitarianism emphasizes on producing the greatest possible balance of positive value for everyone affected. This ethical position determines a decision based on which option produces the greatest total benefit or the lowest total costs (Figure 5.3).

Utilitarianism deals with making a decision based on the greatest total benefit or the lowest total costs. This perspective of the employment of biometrics represents the disclosure of information that could cause risk of material harm to many individuals. It is important to safeguard the information through fingerprint tests, DNA tests and facial recognition systems.

Relativism is the belief that nothing is objectively right or wrong, and that every point of view is valid (Figure 5.4). Under this position, there is not a lot of restrictions because the decision is based on what the person

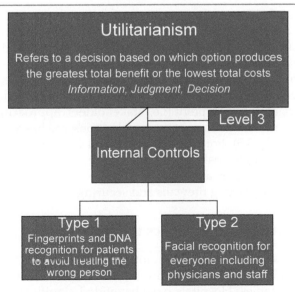

Figure 5.3. Utilitarianism algorithmic system

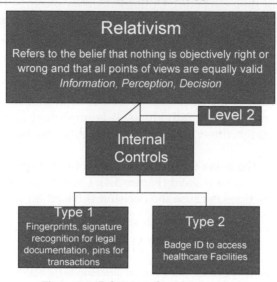

Market Constraints

Recent researchers have estimate that the use of biometrics in the healthcare market will increase the demand for biometrics to be used for healthcare fraud prevention and cost containment in the US.

Figure 5.4. Relativism algorithmic system

believes is valid. This position could be classified as Level 2 because the information disclosed would not cause material harm, but the company has chosen to keep confidential. This level refers to the disclosure of information relative to what the person believes is wrong or right. In this case analysis in the medical field I am proposing implementing the signature recognition to test whether a person is the right individual.

The ethics of care viewpoint (i.e., stakeholders' approach) is a type of ethical consideration that focuses on moral actions and has a more feminist viewpoint (Figure 5.5). For this position, Level 4 would be implemented because it entails more risks and if any of the information. That is why biometrics systems such as facial, gait, and voice recognition are used along with creating strong firewalls and securing databases.

Implementation of Biometrics in Hospitals

Biometric devices are often used for law enforcement purposes but lately they have gained popularity as a means to improve internal control in companies. Hospitals have benefitted significantly from these devices. For this reason, an analysis was made of these controls would

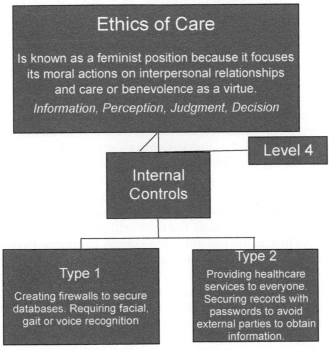

Environment (Nature)

Recent researchers have estimated that the use of biometrics in the healthcare market will increase the demand for biometrics to be used for healthcare fraud prevention and cost containment in the US.

Ethics of Care

Is known as a feminist position because it focuses its moral actions on interpersonal relationships and care or benevolence as a virtue.
Information, Perception, Judgment, Decision

Level 4

Internal Controls

Type 1

Creating firewalls to secure databases. Requiring facial, gait or voice recognition

Type 2

Providing healthcare services to everyone. Securing records with passwords to avoid external parties to obtain information.

Figure 5.5. Ethics of care algorithmic system

benefit Johns Hopkins Hospital (referred from this point forward as "the Company" or "the Organization"). The Company has a very tight control environment which has resulted from both external and internal factors. The external factors affecting the Organization are laws such as the Health Insurance Portability and Accountability Act (HIPAA) which demands health providers to disclose medical records to patients among other things. The internal factors are the structure of the Company which includes audit and ethics committees as well as the vision and mission statements and core values which revolve around integrity and respect and set the tone at the top. Due to the fact that hospitals must operate under very strict rules but at the same time seek the greatest good for the greatest number, an ethical analysis was conducted in which deontology and utilitarianism were compared. Deontology was determined to be the pathway that supported the use of biometric devices better since consequences are irrelevant. It was also chosen as the pathway that hospitals must ultimately follow due to the threat of lawsuits. After this, different biometrics were compared, and the palm scan was chosen as the best for the Organization since it relies on physiological traits which are hard to replicate and is the most cost effective. This biometric was implemented in the revenue, payroll, and fixed assets cycles of the Company to strengthen controls. COSO (Committee of Sponsoring Organizations of the Treadway Commission) is a joint initiative of private sector organizations and is dedicated to issuing thought leadership through the advancement of frameworks and guidance on organization risk management, internal control, and fraud deterrence (https://www.aicpa.org/interestareas/businessindustryandgovernment/resources/riskmanagmentandinternalcontrol/coso-integrated-framework-project.html). Amendment V was determined to be a threat to the use of biometric evidence since it establishes that individuals may not be witnesses against themselves, but the benefits offered by the biometric system outweighed this threat.

Biometric technology has become very appealing to different industries because of the advantages it offers, such as expediting processes and authenticating information. For the healthcare industry, it has become an area of particular interest since the Health Insurance Portability and Accountability Act (HIPAA) was passed in 1996. The HIPAA act offers protection for patients' personal health information, including electronic records. It also sets a limit on the use and release of patient records, establishing privacy standards for health providers, and giving patients the right to request a written notice on how their health information is being used and shared (US Department of Health & Human Services Office for Civil Rights). For this reason, health providers could greatly benefit from the implementation of biometric internal controls. This paper seeks to evaluate the different biometrics available to determine which one would best satisfy the hospital's needs. Also, an analysis will be done of the revenue, payroll, and fixed assets cycles of Johns Hopkins Hospital to determine how the organization could decrease fraud by implementing biometric internal controls.

Company profile

Johns Hopkins Hospital opened its doors in 1889 and The Johns Hopkins University School of Medicine opened four years later. Now known as John Hopkins Medicine, this company combines research, patient care, and education. It is a $5 billion dollar system that brings together doctors, scientists, health professionals and facilities. The mission of the company is, "to improve the health of the community and the world by setting the standard of excellence in medical education, research and clinical care.

Diverse and inclusive, Johns Hopkins Medicine educates medical students, scientists, health care professionals and the public; conducts biomedical research; and provides patient-centered medicine to prevent, diagnose and treat human illness ("Johns Hopkins medicine"). The company refers to its vision as the following: "Johns Hopkins Medicine" provides a diverse and inclusive environment that fosters intellectual discovery, creates and transmits innovative knowledge, improves human health, and provides medical leadership to the world." The core values of the company are: excellence and discovery, leadership and integrity, diversity and inclusion, and respect and collegiality ("Johns Hopkins medicine"). The Company also has a Patient Bill of Rights available that delineates the patients' rights and responsibilities (see Appendix, Item 1).

Analysis of different biometric controls

Many different biometric controls have been developed throughout the years. All the devices have the same ultimate objective, authenticating an individual's identity. However, different devices offer different capabilities. This means that some devices may be inappropriate for a hospital environment. Also, other factors such as price, the time needed to create and run the database, and the enrollment rates need to be taken into consideration when choosing the appropriate biometric device.

There are two main types of biometric devices: physiological and behavioral. Physiological devices gather information from a person's body parts (like the iris or fingerprint) and behavioral from the particular way in which a person does something (like walks or talks). We consider physiological devices to be superior to behavioral because behaviors can be imitated while an individual's physiology is unique and almost impossible to replicate. Because of this way we will only consider physiological devices in our evaluation of biometric techniques. All the biometric devices proposed will be for verification purposes, not identification since it is more cost efficient for the Organization.

A brief description of the characteristics of the following biometric devices are discussed: iris recognition, facial recognition, fingerprint scan, retinal scan, hand geometry, and palm scan. Iris recognition has the highest accuracy. According to the Biometric Product Testing Final Report this device had no false matches in over two million cross-comparisons. Iris recognition can also process large populations at a high speed. Other important benefits are that the iris of a person does not change throughout their lives and individuals agree to being scanned which reduces privacy concerns. Facial recognition offers the ability of being able to scan large crowds and match them to the database. This is particularly useful for surveillance purposes because it does not require user cooperation. One concern with this technology is that individuals' faces change with age and surgery and the device can be easily fooled with masks. Also, privacy issues arise because the individuals being scanned may not have agreed to it.

Fingerprint scanning is mostly used for background checks in law enforcement. One benefit of this device is that multiple fingers can be scanned and enrolled into the database. However, it is not as accurate as iris recognition. The false acceptance rate for fingerprint scan has been calculated to be about 1 in 100,000 while iris recognition only has a rate of 1 in 1.2 million. Also, fingerprint scanners obtain about 40 to 60 characteristics while iris recognition measures about 240 characteristics. Fingerprint scanning also takes longer to match records, may return multiple matches depending on the quality of the device and may be forged. Retinal scan offers similar advantages to the iris scan. First, the retina seldom changes throughout a person's life, verification and identification times are very fast, and these devices capture up to 400 characteristics. However, the price of these devices is very high, some people consider them to be highly invasive, and enrollment times are longer than those for iris scan and fingerprinting (Ravi, 2007).

Hand geometry is easy to use but offers several disadvantages. Several conditions such as age, pregnancy and illness affect hand size and it requires large equipment. The last device is the palm vein scan, this device is very accurate because vein patterns are unique and impossible to replicate. This device has a very low failure to enroll, false rejection and false acceptance rates (Sarkar et al., 2010). It is also important to note, in general, that for all the devices that require the user to touch or come into contact with the device hygiene issues may arise (Biometric comparison guide). Lastly, the approximate prices of the devices are presented in Table 5.1 (according to Fulcrum biometrics and FindTheBest).

Hospitals have a special environment which has to be taken into consideration when selecting the appropriate biometric device. This first thing to consider is the traffic of patients; on busy days a device that has a faster enrollment rate would be preferable. Also, patients sometimes arrive unconscious or injured, so a device that does not require them to be

Table 5.1. Typical price range for biometrics devices

Biometric Device	Approximate Price
Facial recognition	$471.00–6,811.00
Iris recognition	$180.00–2,699.00
Fingerprint scan	$35.99–2,250.00
Retinal recognition	$575.00–3,100.00 (annually)
Hand geometry	$825.00–2,800.00
Palm vein scan	$28.00–399.00

conscious is needed. A palm vein scan is considered the best solution for hospitals for several reasons. First, it is cost-effective, so the hospital can purchase more of these units for the same amount a more expensive device would cost, and in this way improve their customer service. Second, it has a high enrollment rate and lower error rates. Third, unconscious, injured and elderly patients may find this more comfortable or effective. Due to its efficiency, the palm vein scan could also be used with hospital employees for the first two reasons mentioned previously.

The COSO framework

The COSO Framework requires the evaluation of the Company's control environment, risk assessment, control activities, information and communication, and monitoring activities. The Framework will be applied to the Organization in the following section.

Control environment

According to the integrated framework of 2011, control environment refers to the "set of standards, processes, and structures that provide the basis for carrying out internal control across the organization." This alludes directly to the ethical values of the organization, the independence of the Board of Directors and its oversight and control activities, management responsibilities, the talents that the organization recruits and the individual responsibility of internal control ("Internal control-integrated framework," 2011).

The mission and vision statements as well as the core values of the company were previously presented in the company profile section. These core values include excellence, integrity, and respect which set the tone at the top for the organization. Johns Hopkins Medicine has a Board of Trustees in charge of governance and it is important to note that the organization has two insiders on its Board: Edward Miller Jr, M.D. who is the Vice Chairman of the Trustees and Chief Executive Officer of Johns Hopkins Medicine and Ronald Peterson who is Trustee and President of Johns Hopkins Health System ("Company overview of the Johns Hopkins Hospital", 2013). The Executive Committee of the Organization is in turn composed of the following committees: audit, compensation, facilities and real estate, finance, and nominating. Johns Hopkins Medicine also has a House Staff Council which is a, "self-governing body and representative body of the interns and residents of the Johns Hopkins Hospital and the Bayview Campus." This Council has an ethics committee that is responsible for answering ethical questions and educating health care workers on how to solve ethical dilemmas and other ethical issues ("Johns Hopkins medicine").

According to the Health Insurance Portability and Accountability Act (HIPAA), patients have the right to obtain a copy of their medical records and request information to be changed if it is wrong. Even if the information is correct, patients still have the right to note on their file that they disagree with the information in question. Also, patients can request to know who has seen their information (US Department of Health & Human Services Office for Civil Rights). This Act sets transparency standards that require hospitals to improve the quality of their medical records (see Appendix, Item 2).

Ethical analysis

The throughput model will be used to determine if biometric controls are appropriate in the Organization. The previous discussion about HIPAA suggests that the Hospital must follow very strict rules but also, the environment of hospitals requires the decision making model to provide the greatest good for the greatest amount of people. For this reason, the two ethical pathways that will be discussed are the deontological and utilitarianism. Deontology follows the P→J→D model in which information is not considered. Under the perception that rules need to be followed regardless of the result, biometric controls fit nicely in the Organization since they will ensure that no one gains unauthorized access and that the risk of fraud is minimized. On the other hand, utilitarianism follows the I→J→D model in which perception is replaced by information. Under this perspective consequences matter, so the restrictions that biometric controls create may go against the objective of this ethical model since employees will be limited to their duties and will be incapable of performing additional duties for which they are not authorized even if it costs someone's well-being. Since hospitals are so heavily regulated and are the target of many lawsuits, the deontology approach must be followed. Consequently, biometrics are the perfect controls for hospitals.

In short, the ethical standards set by the Organization, the proper oversight established (audit and ethics committee), and the strict laws that HIPAA enacts create a low fraud environment at Johns Hopkins Hospital.

Risk assessment

As per the COSO Framework, a risk is defined as, "the possibility that an event will occur and adversely affect the achievement of objectives." The fraud triangle needs to be analyzed in this area to identify potential risks of fraud and to be able to deter them. Any weaknesses in internal controls provide opportunities for individuals to commit fraud. For this reason biometric internal controls are suggested to strengthen internal controls and ensure that no one gains unauthorized access to critical departments.

A risk assessment of the revenue, payroll and fixed assets cycle will be made to identify the internal control features that can be strengthened with biometric controls and in this way ensure that the opportunity area of the fraud triangle is minimized.

The revenue cycle

The first step for hospitals is to register the patients to create a record. Some of the information that needs to be included in this record is the patients' personal information, their billing information, medical information, account number, and medical record number (Magovern and Jurek, 2009). The internal control issue in this step is that if the hospital has weak controls, identity theft may occur. Thieves may be tempted to steal someone's identity to be able to see a doctor, get prescription drugs, and file claims with the insurance provider (Malida, 2012). The biometric solution could be that when patients are registered, their palm scan should be taken and entered into the system. This way, whenever a patient returns to the hospital, their identity can be established with the palm scan and the chances of identity theft occurring will decrease significantly. This is a preventive control against fraud.

Next, the hospital needs to explain the institution's payment policy. During this step, insurance coverage should be verified. Subsequently, copies of the patient's ID cards and insurance cards should be made along with any additional applicable supporting documentation. During the patient's stay at the hospital, the procedures, treatments, and care that the patient receives need to be recorded. This is typically done using a charge description master (CDM) which is a digital list of codes. These codes include information such as the hospital department, the services charged, a description, a medical code, a standard claim form revenue code, and the dollar amount of the expense. Codes are of crucial importance since they will also be reported on claims for reimbursement, so proper coding will allow for maximum reimbursement. Billing accuracy needs to be reviewed by reconciling the total patient's charges against the CDM and medical record (Magovern and Jurek, 2009). The internal control issue here is that the records may be reconciled by the same person and the signature or initials of the reviewer may be forged. The biometric solution that can be implemented is, again, using a palm scan to verify the identity of the preparer and the reviewer of the reconciliation and in this way ensuring proper segregation of duties. This is a preventive control against fraud since unauthorized users will not be allowed to perform the tasks.

Afterwards, when the patient is discharged, the discharged date needs to be recorded on the system (Magovern and Jurek, 2009). The internal control problem is that a different discharge date may be recorded in order to be able to submit a higher claim for reimbursement ("Common fraud and abuse schemes," 2011). The biometric solution that can prevent this from happening is having the patients check out using the palm scan and preparing a reconciliation of the palm scan records with the information presented in the claims for reimbursement. The controls previously mentioned for reconciliations can also be applied in this area. This is a preventive control against fraud.

After the billings reconciliation, claims may be prepared and transmitted for reimbursement such as Medicare. Bills need to be monitored to ensure that they are paid on time and that any necessary follow-up is made. For patients that have partial or no insurance coverage, statements need to be generated and the proper follow-up also needs to be enforced (Magovern and Jurek, 2009). Billing is the most fraud-prone area in the revenue cycle for hospitals because several types of fraud can occur. One very sensitive area nowadays is Medicare reimbursement. Medicare reimburses for just certain extremely expensive services, which are considered outlier services (Fields, 2010). As such, billings can be increased for expensive services to obtain higher Medicare reimbursements (Abelson and Creswell, 2012). Hospitals received one billion more in Medicare reimbursements in 2010 than they did five years earlier, in part by committing fraud (Abelson and Creswell, 2012). Electronic health records can be coded inappropriately to obtain Medicare payments. An example is cloning records which is used to inflate Medicare reimbursement. Also, hospitals may lie on the health records by saying patients were more ill and needed more care than what they actually needed to get higher reimbursements (Abelson and Creswell, 2012). The biometric solution is that the billings reconciliation needs to be reviewed by at least two people to ensure proper verification of the medical records charged to the bills. Again, identity of the reviewers can be authenticated through a palm scan. This control is mainly preventive but it can also act as detective in that if the records have been manipulated it will allow for their detection.

The payroll cycle

The steps in the payroll cycle will now be discussed. First, the time worked by employees and attendance data needs to be collected. Methods usually used to gather this data are clock cards, time sheets, and time books (Aseervatham and Anandarajah, 2003). The internal control issue is that phantom employees may be used; personnel may clock in their friends even when they did not attend work or employees that have been terminated could continue to appear on the payroll. The biometric solution proposed is that instead of using clock cards, time sheets and time clocks, the hospital should implement a biometric check in, such as a palm scan to authenticate the identity of the person clocking in. This is a preventive control against fraud.

Next, gross wages payable is calculated. This calculation includes regular hours worked, overtime, incentive pay, commissions, and any other allowances. After this figure is obtained, taxes have to be computed and the payroll records need to be updated. The head of the department needs to review the payroll calculation before checks are disbursed to the employees. The internal control weakness is that an unauthorized person could approve the payroll and commit fraud in this way. The biometric solution is that the reviewer should authenticate their identity using the palm scan. This is a preventive control against fraud.

Subsequently, payroll is disbursed to employees usually through checks or electronic bank transfers. Payment summaries need to be generated for each employee by the 14th day after the financial year end or after the employee's termination (Aseervatham and Anandarajah, 2003).

The fixed assets cycle

The fixed assets cycle entails three basic areas: asset acquisition, asset maintenance, and asset disposal. During the asset acquisition stage, a purchase requisition needs to be generated and approved by management (Hall, 2013). The internal control problem in this step is that signature or initials of approver may be forged to purchase unallowable assets with the hospital's funds. Biometric controls such as palm scan can authenticate the identity of the approver and ensure that segregation of duties exists. This is a preventive control against fraud.

Next, a purchase order needs to be created and a copy is sent to the vendor. Upon arrival of the asset, the receiving clerk will prepare a receiving report (Hall, 2013). The control issue is that unauthorized people may enter the receiving department and steal the assets received. Identity authentication should be conducted via a palm scan to ensure that only authorized people enter the department. Also, other methods such as gait could be implemented (if the hospital desires it) to ensure that the people inside the department are authorized to be there. The palm scan would be a preventive control, if the hospital chose to implement gait, this would be a detective control.

Then, the asset will be recorded in the fixed asset subsidiary ledger. Information such as the asset's estimated life, residual value, depreciation method, and location within the Organization need to be recorded. A depreciation schedule is usually prepared and stored after this. This schedule will be used to record end-of-period depreciation automatically. Any maintenance or repairs done to the assets need to be evaluated to determine if they will extend the useful life of the asset. Any changes in the custody of the assets need to be reported and recorded. For disposing of the assets a disposal report needs to be created and authorized (Hall, 2013). Again, unauthorized individuals may authorize the disposal of the assets. Identity authentication should be done with a palm scan to ensure segregation of duties and proper authorization. This is a preventive control against fraud.

The asset will then be removed from the records. Any gain or loss associated with the disposal needs to be properly recorded. A report with the details of the deletion of the assets needs to be reviewed by the fixed assets department (Hall, 2013).

Control activities

Control activities are put in place by management to ensure that the Organization's objectives are achieved. The biometric control activities previously presented in the risk assessment section show how management can ensure tighter controls.

Information and communication

Relevant and reliable information needs to be used for internal and external communication purposes. Using biometric controls helps improve the quality of the information used at Johns Hopkins Hospital in several ways. First, the patient records will be more accurate since the possibility of identity theft will be virtually reduced to zero. Second, the accounting records will also be benefited since segregation of duties will be ensured with biometric authentication.

Monitoring activities

These activities include evaluations to ensure that the other elements in the COSO Framework are functioning as they should. At Johns Hopkins Hospital, the audit committee is in charge of the monitoring activities. The revenue, payroll, and fixed assets cycles of the Company (previously presented in a narrative format in the risk assessment section of the COSO Framework) will now be presented in a flowchart format. The suggested biometric controls have been incorporated in the flowcharts (Figures 5.6–5.8).

Caveats Pertaining to Privacy Rights

In general, biometric techniques are controversial because many people believe they represent a violation of their rights and a breach of personal privacy. Several amendments of the US Bill of Right make biometrics

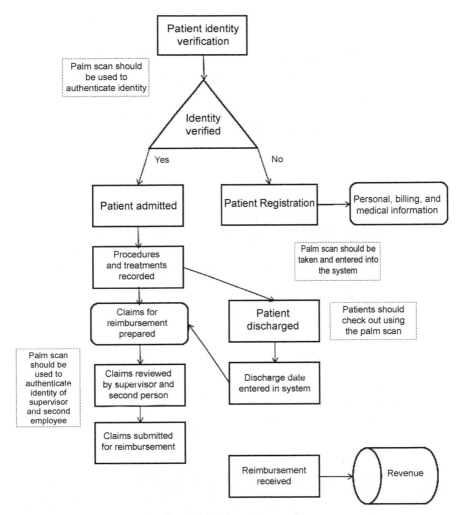

Figure 5.6. The revenue cycle

particularly questionable. The first one is Amendment III which, in a few words, specifies that if someone violates a person's house they are violating their rights. Since biometrics are often used for surveillance purposes, this Amendment makes biometrics very controversial since, technically, procedures like intercepting phone calls to run the voice recognition software would be in violation of this Amendment.

The second Amendment in question is Amendment IV which states, "The right of the people to be secure in their persons, houses, papers, and effects, against unreasonable searches and seizures, shall not be violated, and no warrants shall issue, but upon probable cause, supported by oath

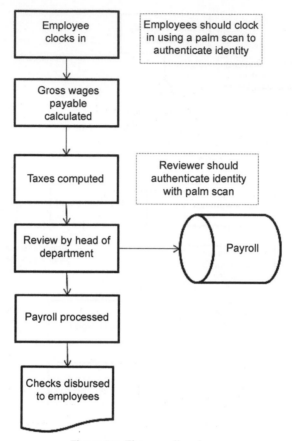

Figure 5.7. The payroll cycle

or affirmation, and particularly describing the place to be searched, and the persons or things to be seized." The surveillance method mentioned before of intercepting phone calls could be considered an unreasonable search and violates this Amendment too.

The last Amendment in question is Amendment V which states that a person cannot be a witness against themselves. Since the uses the hospital would have for the biometric techniques are quite different from those used in law enforcement agencies, we can conclude that Amendments III and IV would not be violated if Johns Hopkins Hospital implements biometric internal controls. However, Amendment V may be of concern in the case of fraud. If fraud is detected in the organization and the biometric records indicate that someone with authority overrode the controls, technically the case could be made that this amendment is being violated since the evidence obtained from the devices comes from the suspect and it cannot be used against the suspect.

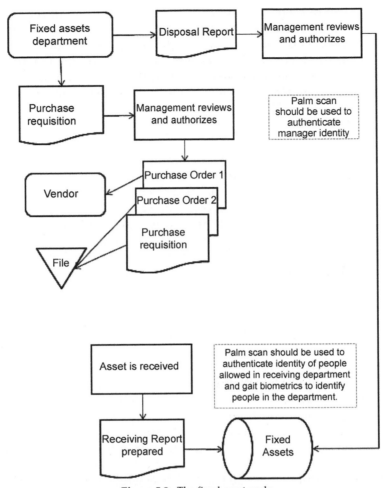

Figure 5.8. The fixed asset cycle

In any case, even if the biometric records may not be acceptable evidence they will point the investigators in the right direction and further investigation can be conducted to obtain additional evidence or a confession. As such, in a hospital environment, the violation of the Bill of Rights is not considered to be a significant issue that may outweigh the benefits of having a biometric control system in place.

Recommendations

In short, the Organization should use palm scan to strengthen internal controls. This device is cost effective so it can be implemented in most of the business cycles of the Company. It is also highly reliable and convenient.

It is important to keep in mind that some patients may be uncomfortable with these devices and have privacy concerns. For this reason, it is crucial to inform patients of the purpose that the device is intended solely for the use of medical records. Patients need to have the option to refuse this service, which may be an inconvenience for the hospital but is necessary since patients cannot be forced to provide their biometric information.

Conclusion

Overall, this chapter highlights the complexity and rapid speed of development in the Artificial Intelligence and ethical considerations area. Governments, businesses and public bodies will need to consider their use of algorithms in decision-making, consulting broadly, and confirming that apparatuses are in place to detect/prevent as well as address any mistakes or unintended consequences of decision choices made.

This century has witnessed the emergence (and dissolution) of new principles and initiatives from a variety of sources, as well as critical public discourse about the development of professional and ethical norms and (the lack of) accountability mechanisms. To sum up, this chapter does not attempt to encapsulate every aspect of the current landscape. Rather, we offer a useful rubric of ethical decision-making *framework* for conceptualizing and hypothesizing about the role of professional norms in the governance of Artificial Intelligence.

The Throughput Model's ethical decision-making framework reflects that the current landscape of professional norms that are likely candidates to display governing effects on Artificial Intelligence-based technologies, largely defined, is fundamentally fragmented and in fluctuation. The body of relevant norms does not present itself in the form of a coherent normative structure, such as a distinct ethics code; however, it is rather a patchwork consisting of existing as well as emerging norms. They can be more general or Artificial Intelligence specific in nature and emerge from the six dominant ethical positions. Ethical considerations can interact with these positions in an organizational setting.

We argue in the chapter that governing effects might stem from the discussed six dominant ethical positions that are *input-oriented* in that they can deal with the circumstances under which Artificial Intelligence-based technologies are created or *output-oriented* by addressing the use of such technologies.

APPENDIX

Item 1: Patient Bill of Rights and Responsibilities
Item 2: Your Health Information Privacy Rights

Patient Bill of Rights and Responsibilities

We want to encourage you, as a patient at The Johns Hopkins Hospital, to speak openly with your health care team, take part in your treatment choices, and promote your own safety by being well informed and involved in your care. Because we want you to think of yourself as a partner in your care, we want you to know your rights as well as your responsibilities during your stay at our hospital. We invite you and your family to join us as active members of your care team.

Your Rights

- YOU HAVE THE RIGHT to receive considerate, respectful and compassionate care in a safe setting regardless of your age, gender, race, national origin, religion, sexual orientation, gender identity or disabilities.

- YOU HAVE THE RIGHT to receive care in a safe environment free from all forms of abuse, neglect, or mistreatment.

- YOU HAVE THE RIGHT to be called by your proper name and to be in an environment that maintains dignity and adds to a positive self-image.

- YOU HAVE THE RIGHT to be told the names of your doctors, nurses, and all health care team members directing and/or providing your care.

- YOU HAVE THE RIGHT to have a family member or person of your choice and your own doctor notified promptly of your admission to the hospital.

- YOU HAVE THE RIGHT to have someone remain with you for emotional support during your hospital stay, unless your visitor's presence compromises your or others' rights, safety or health. You have the right to deny visitation at any time.

- YOU HAVE THE RIGHT to be told by your doctor about your diagnosis and possible prognosis, the benefits and risks of treatment, and the expected outcome of treatment, including unexpected outcomes. You have the right to give written informed consent before any non-emergency procedure begins.

- YOU HAVE THE RIGHT to have your pain assessed and to be involved in decisions about treating your pain.

- YOU HAVE THE RIGHT to be free from restraints and seclusion in any form that is not medically required.

- YOU CAN EXPECT full consideration of your privacy and confidentiality in care discussions, exams, and treatments. You may ask for an escort during any type of exam.

- YOU HAVE THE RIGHT to access protective and advocacy services in cases of abuse or neglect. The hospital will provide a list of these resources.

- YOU, YOUR FAMILY, AND FRIENDS WITH YOUR PERMISSION, HAVE THE RIGHT to participate in decisions about your care, your treatment, and services provided, including the right to refuse treatment to the extent permitted by law. If you leave the hospital against the advice of your doctor, the hospital and doctors will not be responsible for any medical consequences that may occur.

- YOU HAVE THE RIGHT to agree or refuse to take part in medical research studies. You may withdraw from a study at any time without impacting your access to standard care.

- YOU HAVE THE RIGHT to communication that you can understand. The hospital will provide sign language and foreign language interpreters as needed at no cost. Information given will be appropriate to your age, understanding, and language. If you have vision, speech, hearing, and/or other impairments, you will receive additional aids to ensure your care needs are met.

- YOU HAVE THE RIGHT to make an advance directive and appoint someone to make health care decisions for you if you are unable. If you do not have an advance directive, we can provide you with information and help you complete one.

- YOU HAVE THE RIGHT to be involved in your discharge plan. You can expect to be told in a timely manner of your discharge, transfer to another facility, or transfer to another level of care. Before your discharge, you can expect to receive information about follow-up care that you may need.

Obtained from the Johns Hopkins Hospital website

- You HAVE THE RIGHT to receive detailed information about your hospital and physician charges.

- You CAN EXPECT that all communication and records about your care are confidential, unless disclosure is permitted by law. You have the right to see or get a copy of your medical records. You may add information to your medical record by contacting the Medical Records Department. You have the right to request a list of people to whom your personal health information was disclosed.

- You HAVE THE RIGHT to give or refuse consent for recordings, photographs, films, or other images to be produced or used for internal or external purposes other than identification, diagnosis, or treatment. You have the right to withdraw consent up until a reasonable time before the item is used.

- IF YOU OR A FAMILY MEMBER NEEDS TO DISCUSS an ethical issue related to your care, a member of the Ethics Service is available by pager at all times. To reach a member, dial 410-283-6104. After three beeps, enter your phone number and then the pound sign (#). An Ethics Service member will return your call.

- You HAVE THE RIGHT to spiritual services. Chaplains are available to help you directly or to contact your own clergy. You can reach a chaplain at 410-955-5842 between 8am and 5pm weekdays. At other times, please ask your nurse to contact the chaplain on call.

- You HAVE THE RIGHT to voice your concerns about the care you receive. If you have a problem or complaint, you may talk with your doctor, nurse manager, or a department manager. You may also contact the Patient Relations Department at 410-955-2273 or email patientrelations@jhmi.edu.

If your concern is not resolved to your liking, you may also contact:

Maryland Department of Health & Hygiene
Office of Health Care Quality
Hospital Complaint Unit
Spring Grove Hospital Center
Bland Bryant Building
Catonsville, Maryland 21228
410-402-8000

The Joint Commission
Office of Quality Monitoring
One Renaissance Boulevard
Oakbrook Terrace, IL 60181
1-800-994-6610
complaint@jointcommission.org

Your Responsibilities

- You ARE EXPECTED to provide complete and accurate information, including your full name, address, home telephone number, date of birth, Social Security number, insurance carrier and employer when it is required.

- You SHOULD PROVIDE the hospital or your doctor with a copy of your advance directive if you have one.

- You ARE EXPECTED to provide complete and accurate information about your health and medical history, including present condition, past illnesses, hospital stays, medicines, vitamins, herbal products, and any other matters that pertain to your health, including perceived safety risks.

- You ARE EXPECTED to ask questions when you do not understand information or instructions. If you believe you cannot follow through with your treatment plan, you are responsible for telling your doctor. You are responsible for outcomes if you do not follow the care, treatment, and service plan.

- You ARE EXPECTED to actively participate in your pain management plan and to keep your doctors and nurses informed of the effectiveness of your treatment.

- You ARE ASKED to please leave valuables at home and bring only necessary items for your hospital stay.

- You ARE EXPECTED to treat all hospital staff, other patients, and visitors with courtesy and respect; abide by all hospital rules and safety regulations; and be mindful of noise levels, privacy, and number of visitors.

- You ARE EXPECTED to provide complete and accurate information about your health insurance coverage and to pay your bills in a timely manner.

- You HAVE THE RESPONSIBILITY to keep appointments, be on time, and call your health care provider if you cannot keep your appointments.

OCTOBER 2012

Obtained from the Johns Hopkins Hospital website

YOUR HEALTH INFORMATION PRIVACY RIGHTS

Most of us feel that our health information is private and should be protected. That is why there is a federal law that sets rules for health care providers and health insurance companies about who can look at and receive our health information. This law, called the Health Insurance Portability and Accountability Act of 1996 (HIPAA), gives you rights over your health information, including the right to get a copy of your information, make sure it is correct, and know who has seen it.

Get It.

You can ask to see or get a copy of your medical record and other health information. If you want a copy, you may have to put your request in writing and pay for the cost of copying and mailing. In most cases, your copies must be given to you within 30 days.

Check It.

You can ask to change any wrong information in your file or add information to your file if you think something is missing or incomplete. For example, if you and your hospital agree that your file has the wrong result for a test, the hospital must change it. Even if the hospital believes the test result is correct, you still have the right to have your disagreement noted in your file. In most cases, the file should be updated within 60 days.

Know Who Has Seen It.

By law, your health information can be used and shared for specific reasons not directly related to your care, like making sure doctors give good care, making sure nursing homes are clean and safe, reporting when the flu is in your area, or reporting as required by state or federal law. In many of these cases, you can find out who has seen your health information. You can:

- **Learn how your health information is used and shared by your doctor or health insurer.** Generally, your health information cannot be used for purposes not directly related to your care without your permission. For example, your doctor cannot give it to your employer, or share it for things like marketing and advertising, without your written authorization. You probably received a notice telling you how your health information may be used on your first visit to a new health care provider or when you got new health insurance, but you can ask for another copy anytime.

- **Let your providers or health insurance companies know if there is information you do not want to share.** You can ask that your health information not be shared with certain people, groups, or companies. If you go to a clinic, for example, you can ask the doctor not to share your medical records with other doctors or nurses at the clinic. You can ask for other kinds of restrictions, but they do not always have to agree to do what you ask, particularly if it could affect your care. Finally, you can also ask your health care provider or pharmacy not to tell your health insurance company about care you receive or drugs you take, if you pay for the care or drugs in full and the provider or pharmacy does not need to get paid by your insurance company.

Obtained from the Health and Human Services website

References

Abelson, R. and Creswell, J. (2012, September 24). U.S. warning to hospitals on Medicare bill abuse. The New York Times. Retrieved from http://www.nytimes.com/2012/09/25/business/us-warns-hospitals-on-medicare-billing.html?_r=0.

Aseervatham, A. and Anandarajah, D. 2003. Accounting information & reporting systems. Australia: McGraw-Hill. Retrieved from highered.mcgrawhill.com/sites/dl/free/0074711407/../ppt_ch14.ppt.

Biometric comparison guide. In iridian technologies. Retrieved from http://epic.org/privacy/surveillance/spotlight/1005/irid_guide.pdf.

Common fraud and abuse schemes. (2011). Retrieved from http://www.aging.ks.gov/SHICK/Fraud_Abuse/Fraud_Abuse_Schemes.htm.

Fields, R. 2010. 15 fraud and abuse cases making headlines in 2010. *Becker's Hospital Review*, Retrieved from http://www.beckershospitalreview.com/hospital-management-administration/15-fraud-and-abuse-cases-making-headlines-in-2010.html.

Hall, J. 2013. Accounting information systems (8th ed., p. 278). Mason: South-Western, Cengage Learning.

Hastie, T., Tibshirani, R. and Friedman, J. 2001. The Elements of Statistical Learning: Data Mining, Inference, and Prediction. 1st edn. Springer Series in Statistics. New York: Springer.

Internal Control Integrated Framework. 2011. Committee of Sponsoring Organizations of the Treadway Commission.

John Hopkins Hospital. 2013. Company overview. Retrieved from http://investing.businessweek.com/research/stocks/private/board.asp?privcapId=4183014.

Johns Hopkins Medicine. (n.d.). Retrieved from http://www.hopkinsmedicine.org.

Magovern, S. and Jurek, J. 2009. Hospital Billing (2nd edn.). McGraw-Hill Higher Education. Retrieved from http://highered.mcgraw-hill.com/sites/0073520896/student_view0/.

Malida, J. 2012. The best offense is a good defense when dealing with electronic medical records. Employee Benefit Adviser, 10(2): 12–14.

Ravi, D. 2007. Retinal recognition biometric technology in practice. Keesing Journal of Documents & Identity, (22), Retrieved from http://www.biometricnews.net/Publications/Biometrics_Article_Retinal_Recognition.pdf.

Rodgers, W. and Gago, S. 2001. Cultural and ethical effects on managerial decisions: Examined in a Throughput Model. Journal of Business Ethics, 31: 355–367.

Rodgers, W. and Gago, S. 2003. A model capturing ethics and executive compensation. Journal of Business Ethics, 48: 189–202.

Rodgers, W. and Gago, S. 2004. Stakeholder influence on corporate strategies over time. Journal of Business Ethics, 52: 349–363.

Rodgers, W. 2006. Process Thinking: Six Pathways to Successful Decision Making. NY: iUniverse, Inc.

Rodgers, W. 2009. Ethical Beginnings: Preferences, Rules, and Principles Influencing Decision Making. NY: iUniverse, Inc.

Rodgers, W., Guiral, A. and Gonzalo, J.A. 2009. Different pathways that suggest whether auditors' going concern opinions are ethically based. Journal of Business Ethics, 86(2009): 347–361.

Rodgers, W., Söderbom, A. and Guiral, A. 2014. Corporate social responsibility enhanced control systems reducing the likelihood of fraud. Journal of Business Ethics, 131(4): 871–882.

Rodgers, W. 2019. Trust Throughput Modeling Pathways. Hauppauge, NY: Nova Publication.

Rodgers, W. and Al Fayi, S. 2019. Ethical Pathways of Internal Audit Reporting Lines. Accounting Forum, 43(2): 220–245.

Rosso, C. 2015. The conundrum of machine learning and cognitive biases. Medium. Pyschology Today, July 14: 3.

Sarkar, I., Alisherov, F., Kim, T. and Bhattacharyya, D. 2010. Palm vein authentication system: A review. International Journal of Control and Automation, 3(1): 27–34. Retrieved from http://www.sersc.org/journals/IJCA/vol3_no1/3.pdf.

U.S. Department of Health & Human Services Office for Civil Rights (n.d.). Your health information privacy rights. Retrieved from http://www.hhs.gov/ocr/privacy/hipaa/ understanding/consumers/consumer_rights.pdf.

6

Cyber Securities Issues

Fraud and Corruption

"Adversarial machine learning' is key area the NITRD cybersecurity R&D strategic plan, that evaluates the extent to which Artificial Intelligence systems can be contaminated by training data, modified algorithms, etc."

—'The National Artificial Intelligence R&D Strategic Plan (Oct 2016)', National Science and Technology Council, USA

"Cybercrime is the greatest threat to every profession, every industry, every company in the world."

—IBM President and CEO Ginni Rometty

Artificial Intelligence continues the movement towards advancement of technology that has unbridled the practical applications, many of which can augment the decision-making process. Algorithms power Artificial Intelligence, and algorithms are driven by large amounts of data. Moreover, Artificial Intelligence, machine learning and deep learning are motivating transformation across essentially all industries and disciplines. They are helping organizations restructure internal processes to improve efficiency, make sense of the vast amounts of data to drive judicious decision-making, and design up-to-date, innovative services to enhance users' experience. A safe and secure Artificial Intelligence system is one that acts in a controlled and well-recognized manner. The design philosophy must be such that it ensures security against external attacks, anomalies and cyberattacks.

Despite the endeavors of authorities around the globe, corrupt and fraudulent behavior continues to flourish, with an incapacitating economic effect. For example, data quality can be affected by a number of aspects that include timeliness, granularity, the quality of metadata and the possibility of calibration error (Royal Academy of Engineering, 2017). There are supplementary concerns about social exclusion and unconscious bias being embedded in these systems. Moreover, privacy issues are important matters such as the threat of 'spoofing', whereby data is falsified intentionally without the knowledge of the data recipient. Any uncertainty may be amplified by the combination of various datasets.

Hence, issues of governance and fraud reduction will need to be considered in the design and development of these systems so that incorrect assumptions about the behavior of users or designers are circumvented.

Fraud is one of the oldest type of crimes in human history. Fraudulent individuals have been there throughout the ages and throughout ages there are people who have been victims of fraud. For example, credit cards are well-known targets of fraudulent activities. With the development of e-marketing, the count of fraudulent activities is on the rise. Users' credit cards information stored in some organizations' databases, such entities types as banks, online shopping businesses or online service providers. There are an increasing instance of frauds on online transactions with the pervasive use of the internet. Because of this, the need for automatic systems that are able to detect and fight fraud has emerged.

In addition, a number of factors has caused the rising in cases of fraud, with technology carrying on producing new avenues. Malicious actors, both internally and externally, are seizing the opportunities created by the digital revolution. Organizations are amassing immeasurable amounts of data, much of which is personal or other confidential information from customers and other users. In the wrong hands, this information may be implemented to commit fraud and other economic crimes.

Data protection safeguards are a necessary condition to be built into software and services from the earliest stages of development. Requirements for systems with properties that can be checked by regulators or the public without compromising data protection is an additional audit check. Procedures could embrace the disclosure of certain key pieces of information, including aggregate results and benchmarks, when communicating algorithmic performance to the public. Further research into applicable apparatuses and strong leadership is vital to address the evolving intellectual property and legal constraints.

The remaining sections of this chapter will discuss: (1) breaching biometric data, (2) the need for machine learning, (3) fraud triangle and Artificial Intelligence, and (4) Advanced Intelligence Stores.

How Biometric Data Can be Breached

If a hacker desires to access a system that is shielded by a fingerprint or face scanner, there are a number of ways a fraudster could accomplish this dastardly task. First, a person's fingerprint or facial scan (template data) is stored in a database that could be replaced by a hacker to gain illicit access to a system.

Second, a physical copy or spoof of a person's fingerprint or face could be produced from a stored template data to gain unapproved access to a system embezzled template data. Also, it could be reused to gain unauthorized access to a system pilfered template data that has been implemented by a fraudster to illicitly track a person from one system to another.

Biometric data need urgent protection

Nowadays, biometric systems are progressively more utilized in government civil, commercial and national defence applications. In addition, consumer devices equipped with biometric systems are found in everyday electronic devices like smartphones. MasterCard and Visa cards both of which come with embedded fingerprint scans. What is more, wearable fitness devices are increasingly using biometrics to unlock smart cars and smart homes.

The Internet of Things, gadgets from thermostats to ovens have become smarter, utilizing sensors, data, and cloud computing to set the ideal temperature or cook an apple pie to a person's personal taste. Devices such as Amazon Echo or Google Home, with their voice assistants, have developed to all intents and purposes a hub of these embryonic autonomous systems.

Organizations produce and collect a large volume of data which may be structured or unstructured in order to enhance their services. Nevertheless, the disorganized data is left unused owing to a drought of advanced technologies. The unstructured data may or may not contain confidential information. Consequently, there is a need to make safe all of the accumulated digital assets. There is always a risk of hackers waiting for an opportunity to steal the classified information. Correspondingly, the Security Intelligence acknowledges that "the global cost of cybercrime will reach $2 trillion by 2019, a threefold increase from the 2015 estimate of $500 billion" (https://securityintelligence.com/20-eye-opening-cybercrime-statistics/).

More examples include the British insurance company Lloyd's, whose assessments are that cyber-attacks cost organizations as much as $400 billion a year. These costs include direct damage plus post-attack disruption to the normal course of operations. Moreover, various vendor and media forecasts put the cybercrime total as high as $500 billion and more (https://www.forbes.com/sites/stevemorgan/2015/11/24/ibms-ceo-on-hackers-cyber-crime-is-the-greatest-threat-to-every-company-in-the-world/#2817919573f0).

In addition, the World Economic Forum claims that a significant part of cybercrime remains concealed, particularly industrial espionage where access to confidential documents and data is difficult to identify by organizations. Although analysts work on cybercrimes, there is a need for machines that can perform smarter and assist analysts to operate systems accurately. Machine learning for cybersecurity may prove valuable to organizations in fortifying their data.

Security and the Dark Web

The Dark Web is the component of the internet that cannot be found through a typical "Google" or "Yahoo" search. It necessitates the implementation of a special browser such as Tor which keeps the user (relatively) anonymous. As a result, it presents a unique challenge for law enforcement. On the Dark Web, there is a wide range of illegal activity going on such as trafficking in stolen goods, illicit substances or weapons, killers for hire, and other despicable transactions.

Many people think of the Dark Web as an epicenter of criminal activity with little reason for a law-abiding citizen to visit it. Nonetheless, while this is often the case, not all activities on the Dark Web are illegal. For example, in countries where internet access is regulated and monitored by the government, the Dark Web can assist people to communicate without fear of repercussions.

Moreover, those concerned with their personal privacy on the internet can find a great deal of advice and different techniques to incorporate it into their lives. In addition, it can be a safe place for whistleblowers to share information. That is, some mainstream media organizations monitor Dark Web whistleblower sites, including a version of Wikileaks. Lastly, law enforcement can benefit from monitoring the Dark Web as part of a greater threat analysis and situational awareness and responsiveness strategy.

Correct software can access the Dark Web. The popular software for it includes Tor, Freenet and I2P (https://null-byte.wonderhowto. com/news/anonymity-networks-dont-use-one-use-all-them-0133881/). Each anonymity network is configured for a diverse specific purpose. A particular network alone cannot perform what the three can jointly accomplish. Tor and I2P cannot persist information like Freenet can, Tor and Freenet cannot provide the generic transports that I2P provides and Freenet doesn't handle data streaming as well as Tor and I2P. Presently, Tor network represents the best proxy system.

Dark Web intelligence can be important for security decision-making. That is, it is possible to collect exploits, vulnerabilities and other indicators of compromise, as well as insight into the techniques, tactics and procedures that cyber crooks utilize for distinct knowledge about the tools and malware which threat actors prefer to employ.

When this real-time threat data is screened through satisfactory context and separated from false positives, it may become actionable intelligence. This may lead to understanding emerging threat trends to develop mitigation techniques proactively. Dark-source intelligence may also assist with ascertaining criminal motivations and collusion prior to attacks. In addition, it may assist in attributing risks and attacks to specific crime groups.

The Need for Machine Learning

For an organization protecting its data, network security is very important, and even small data centers might have hundreds of applications running, each of which requires to have diverse security policies enforced. Individual experts might take many days to weeks to fully comprehend these policies and accomplish that the security implementation has been successful.

Cyber security intrinsically involves repetitiveness and tediousness. This is due to the fact that identification and assessment of cyber threats necessitate sifting through huge amounts of data and searching for anomalous data points. Organizations can utilize the data gathered by their prevailing rules-based network security software to train Artificial Intelligence machine learning algorithms towards pinpointing new cyber threats.

Organizations hire analysts who can work to detect any malicious activities in their network. Since fraudsters are waiting for the right opportunity to spoil the system by using new technologies, organizations must understand benefits that machine learning provides and apply them in their respective sectors. In addition, the traditional cyber security methods and analysts are unsuccessful to address the following problems:

1. As an enormous volume of data streams into organizations, it becomes mind-numbing for analysts to analyze and establish where exactly the malware has been injected.
2. The second concern is an unqualified analyst who is oblivious of the 'what and how' of all-inclusive network security processes.
3. The third issue is even though malware is detected in a network, the additional processes, such as communicating with administrators takes a lot of time. Therefore, it is generally a long drawn process.

The aforementioned reasons spawned the need for more advanced and capable technologies that could support analysts to detect malware and safeguard their systems.

Machine learning for cyber security

As discussed in previous chapters, machine learning is a type of Artificial Intelligence, which gathers large chunks of data and sometimes with the assistance of neural networks and deep learning trains machines to make devices act smarter and as intelligent as humans. Machine learning can assist organizations for cybersecurity in the following ways (https://www.allerin.com/blog/think-cybersecurity-think-machine-learning):

1) Machine learning has an interesting feature, termed predictive analysis, which can predict the future outcomes of any system. Based on the datasets, machine learning can calculate in a probabilistic sense, where, how, and when a hacker will place a malware. This can serve as an early warning system to organizations and alert them to be ready for any unwanted kind of illegal intrusions.

2) Even if fraudsters hack the system, machine learning can collect the experience as datasets and train systems in order to detect comparable malicious activity in the future.

3) The issue of discovering where hackers have placed the malware is deciphered with machine learning. Machine learning can search the entire system and the network and find the malware in a very expedient manner.

4) Another aspect of machine learning support system is providing recommendations. Similar to an Amazon's recommendation for books, machine learning can recommend an organization with actions that could be helpful for securing their system. Machine learning algorithms unearth patterns in data and gain insights pertaining to it, once the networks are trained with datasets.

When it comes to cyber security solutions, Artificial Intelligence infers the use of machine learning techniques that enable computers to learn from data in a manner similar to people.

It therefore, serves as a platform to secure all digital assets efficiently.

Machine learning in cyber security domain

Machine learning can assist in solving the most common tasks including regression, prediction, and classification. In the era of enormously large amounts of data and cyber security talent shortage, machine learning aided by neural networks and deep learning appears to be a very good solution.

Fraud detection is notably a challenging task due to the following:

1) Fraud strategies change time to time, as well as customers' spending habits evolve.

2) Only a few examples of frauds are available; therefore, it is difficult to create a model of fraudulent behavior.
3) Not all frauds are reported, or they are reported long after the event.
4) A small number of transactions only can be timely investigated.

Therefore, with the large number of transactions occurring everyday, organizations:

a) Cannot ask human analyst to check each and every transaction.
b) Hope to automatize to detect fraudulent transaction.
c) Demand accurate prediction, i.e., minimize missed frauds and false alarms.

Nonetheless, machine learning can assist in overcoming the above hazardous situations with systems which developed with machine learning tools. Systems can learn complex fraudulent patterns by examining the data in large volumes. Further, systems can also create optimal models for fraudulent activities which have complex shapes and hence, successfully suggest what can be done for new types of fraud. Furthermore, a machine learning system can adopt itself to timely changing distribution (fraud evolution). Nonetheless, machine learning systems require enough samples to achieve successful learning.

A machine learning approach to Artificial Intelligence uses a system that is capable of learning from experience. It is intended not only for Artificial Intelligence goals (e.g., replicating human behavior) but it can also reduce the efforts and/or time spent for both simple and difficult tasks such as stock price prediction, etc. In other words, machine learning is a system that can recognize patterns by using examples rather than by programing them. Therefore, a system that learns constantly, makes decisions as a function of data rather than algorithms, and change its behavior, is what is machine learning.

A deep learning approach is a set of *techniques* for implementing machine learning that recognizes pattern of patterns like image recognition. The system identifies primarily object edges, a structure, an object type, and then an object itself. The point is that deep learning is not exactly neural networks, which is basically pattern recognition.

The definitions show that cybersecurity field refers mostly to machine learning (not to Artificial Intelligence). And a large part of the machine learning tasks is not human related.

Machine learning means solving certain tasks with the use of an approach and particular methods based on data that an organization employs in its system (Figure 6.1). Therefore, machine learning employs data and uses tools in data science. From Figure 6.1, it is seen data science is the study of data (i.e., features). It comprises emergent methods of

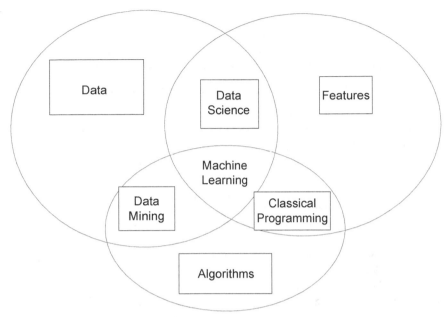

Figure 6.1. Machine learning and components

recording, storing, and analyzing data to successfully extract useful information (i.e., data mining). Moreover, the goal of data science is to obtain insights and knowledge from any type of data, which can be both structured and unstructured for classical programing and algorithms.

The classical programs are the tools that are implemented in machine learning. They can be described as follows:

a) Regression (or prediction)—a task of predicting the next value based on the previous values.
b) Classification—a task of separating things into different categories.
c) Clustering—it is similar to classification but the classes are unknown, and involves grouping things by their similarity.
d) Association rule learning (or recommendation)—a task of recommending something based on previous experience.
e) Dimensionality reduction—or generalization, a task of searching common and most important features in multiple examples.
f) Generative models—a task of creating something based on the previous knowledge of the distribution.

For machine learning approach, a user profile is created for every user in the detection systems. This profile must be updated time to time. When the system has trained with enough samples, it has in-depth information about, for instance, users' spending habits daily, weekly, or at monthly intervals. For example, assume that while a student can afford to spend $200 a week, a business man can spend $2000 per week. While a fraudulent activity with $600 spend at once has high fraud probability for a student, comparably it has very small fraud probability for a business person. Nevertheless, special days such as Thanksgiving, birthdays or weekends must be contemplated when developing an algorithm for fraud detection, since students spend more money on these occasions. Typically, a "fraudster" does not know a victim's spending habits; therefore, fraudulent activity has an inappropriate matching to a particular user profile. Nonetheless, if fraudulent activity matches a user profile, it may be difficult to detect.

As discussed in Chapter 1, regarding the machine learning literature, fraud detection systems can be built with supervised, unsupervised or semi-supervised approaches. Every kind of approach has a slight difference according to working logic.

In supervised learning, historical fraud data is produced in order to develop users' profiles in the training stage. Systems which are trained with unsupervised algorithms can detect unknown fraudulent activities. To minimize the disadvantages of both techniques, semi-supervisory learning approach can be of benefit to an organization. In this approach, supervised and unsupervised algorithms work together, whereby both well-known and unknown fraudulent activities can be detected efficiently.

To sum up, organizations' prime challenges are to minimize the number of false positives being generated, thereby saving time, money, and needlessly frustrating users.

Accurate data analysis

One of the most essential characteristics of machine learning algorithms is that they are able to scrutinize large amounts of transaction data and flag dubious transactions with extremely accurate risk scores in real-time. This risk-based analytics approach detects complex patterns that are challenging for analysts to identify. This includes, but is not limited to, banks and financial organizations, which are far more operationally efficient while detecting more fraud.

The algorithms take into account several factors, including the customer's location, the device implemented, and other contextual data points, to construct a meticulous picture of each transaction. This method enhances real-time decisions and better safeguards customers against fraud, all without influencing the user experience.

This trend will persist in the coming years. Further, with considerable technological growth in this area, organizations will more and more depend on machine learning algorithms to determine which transactions are suspicious.

Fraud Triangle and Artificial Intelligence

If organizations are to proactively combat fraud, they must consider the 'fraud triangle'. This theory was proposed in the 1950s by criminologist Donald R. Cressey. The fraud triangle is a structure conceived to illuminate the reason behind a person's decision to commit fraud. The three stages categorized by the effect on the individual can be summarized as: some degree of pressure (i.e., incentive or motive), some perceived opportunity (i.e., weak internal control system) and some way to rationalize (i.e., a person state of mind: attitude and/or personality) the fraud as not being inconsistent with the individual's values.

Artificial Intelligence concerned with data analysis (e.g., email analysis), with natural language processing (NLP), machine learning and deep learning can scrutinize message content and detect patterns representative of a fraud attempt. These tools can make use of cognitive biases and human vulnerabilities in order to create a context of trust, importance and responsibility that will lead their target to respond positively to their inquiries. Thus, these tools encapsulate the fraud triangle aspects (Rodgers, 2012; Rodgers et al., 2014). The aspects include pressure (i.e., incentive or motive), opportunity, and rationalization. That is, the fraud triangle initiates with an incentive (typically a pressure to do it such as greed, corruption, etc.), followed by an opportunity (i.e., lack of effective cyber internal control), and finally a process of internal rationalization (i.e., state of mind of the fraudster). Since all three of these drivers typically are present for an act of fraud to occur, each of them should be addressed individually.

The perpetrator typically designs an effective attack driven by an *incentive or motive*, which may take time in order to develop a credible plan. The perpetrator identifies their targets that may include social engineering methods (*opportunity*). For example, they may establish initial contact through the internet, email or telephone with the target organization to obtain information that will be implemented during the next stages of their operation. The *rationalization* (i.e., state of mind) process involves identifying someone within an organization who has authority over some transactions of value or potential value. For example, the perpetrator must identify a favorable transactional context, an invoice being processed with a supplier or transfer of funds.

The fraud triangle starts with an incentive (generally a pressure to perform from within the organization) followed by an opportunity, and finally a process of internal rationalization (Figure 6.2). Since all three of these drivers must be present for an act of fraud to occur, each of them should be addressed individually.

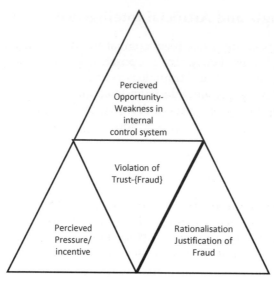

Figure 6.2. Fraud triangle-unfolding the gateway to fraud/cyber-attacks. Source: Rodgers et al., 2014

Pressure (incentive or motive) can come in the form of financial difficulties, debt, or simple greed. These elements can create an incentive (or push factor) to commit fraud.

Rationalization ensues when a perpetrator gives good reason for fraudulent behavior by minimizing or making themselves "feel better" about committing the act. Fraud perpetrators often view committing fraud as a means to recuperate expenditures. That is, they feel they owed payback for the amounts they have given to a person, place or thing. *Opportunity* is the one area that an organization can deploy to disrupt the elements in the triangle and help diminish fraudulent acts.

Blockchain and security against fraud

Blockchain represents a growing list of records, called *blocks*, that are linked using cryptography. By design, a blockchain is impervious to modification of the data. Further, blockchain depicts an open, distributed ledger that can record transactions between two participants efficiently and in a verifiable and permanent manner as well as reducing fraud (Iansiti and Lakhani, 2017). The ledger can also be programed to generate transactions automatically. This tool is at the core of bitcoin and other virtual currencies.

Blockchain was created by an individual (or group of people) making use of the name Satoshi Nakamoto in 2008 in order to operate as the public transaction ledger of the cryptocurrency bitcoin (Prathyusha et al., 2018). The identity of Satoshi Nakamoto is unknown. The invention of the blockchain for bitcoin made it the first digital currency to solve the double-spending problem without the assistant of a trusted authority or central server.

As a security measure, blockchain technology is quite robust since it stores blocks of information that are the same across its network. Moreover, the blockchain cannot be controlled by just one unit, and it has no single point of failure.

Transparent and incorruptible

The blockchain network lives in a state of consensus, one that repeatedly checks in with itself every ten minutes. A type of self-auditing ecosystem of a digital value, the network brings together every transaction that happens in ten-minute intervals. Each group of these transactions is denoted as a "block."

Across global supply chains, financial services, healthcare, government and many other industries, innovators are exploring ways to implement blockchain to unsettle and transform traditional organizational models. Many organizational leaders have accomplished improved business benefits, embracing better transparency, enhanced security, improved traceability, increased efficiency and speed of transactions, and reduced costs. Therefore, five important properties result from this:

1. *Transparency* data is entrenched within the network as a whole, by definition it is public.

   Transaction histories are getting more transparent through the use of blockchain technology. Since blockchain is a kind of distributed ledger, all network participants share the same documentation as compared to individual copies. This particular shared version can only be updated through consensus. To change a single transaction record would necessitate the modification of all prior records and the collusion of the complete network. Therefore, data on a blockchain is more precise, consistent and transparent than when it is rammed through paper-heavy processes. This system is also accessible to all partakers who have authorized access. To change a single transaction record would necessitate the modification of all ensuing records and the collusion of the complete network.

2. *It cannot be corrupted* or change any unit of information on the blockchain, which would indicate implementing a large amount of computing power to override the complete network.

   There are more than a few ways blockchain is more protected than other record-keeping systems. Transactions must be approved before they are recorded. After a transaction is established, it is encrypted and interconnected to the prior transaction. This, along with the understanding that information is warehoused across a network of computers as opposed to on a single server, makes it very challenging for hackers to compromise the transaction data.

3. *Improved traceability*

If your company deals with products that are traded through a complex supply chain, you are familiar with how hard it can be to trace an item back to its origin. When exchanges of goods are recorded on a blockchain, an organization develops an audit trail that displays where an asset came from and every stop it made on its voyage. This historical transaction data can assist in confirming the authenticity of the assets and prevent fraud.

4. *Increased efficiency and speed*

Traditional paper-heavy processes can be very time-consuming and thus prone to human error and habitually requires third-party mediation. By restructuring and automating these processes with blockchain, transactions can be completed quicker and more efficient. Since record-keeping is accomplished utilizing a single digital ledger that is shared among participants, reconciling multiple ledgers is not necessary. Further, when people have access to the same information, it becomes simpler to trust each other without the prerequisite for quite a few intermediaries. Hence, clearing and settlement can happen much faster.

5. *Reduced costs*

For most organizations, shrinking costs is a priority. Blockchain does not require many third parties or distributers to make guarantees since it does not matter if the trading partner can be trusted. As an alternative, one only has to trust the data on the blockchain. Reviewing of documentation is reduced to complete a trade since one and all will have authorization to access a single and immutable version.

The next section deals with a case that employs biometric tools in order to enhance its organization's control systems as well as to diminish fraud.

Case Study of Advanced Intelligence Stores, Inc.

Organizations must rise to this challenge or be disadvantaged relative to traditional and non-traditional competitors. Artificial Intelligence, biometrics, and analytics are central to the success of an enterprise and will pervade critical business areas, including big data, business processes, the workforce, and risk and reputation.

The vision for Artificial Intelligence enhanced biometrics should be guided by innovative thinking—with the long-term objective of enriched, or novel, business strategies and models.

Advanced Intelligence is subject to regulations and laws, such as the Sarbanes-Oxley Act of 2002. The Sarbanes-Oxley Act strictly states and regulates the importance of public company's internal controls. Therefore, this case analyzes Advanced Intelligence's cash receipts, cash disbursements and inventory accounting cycles and internal controls. Also, Advanced Intelligence's intangible assets were examined in this case.

A fraud analysis was conducted to identify potential weaknesses on the accounting cycles. The main weaknesses found in the accounting cycles were authorization and protection of assets issues. In order to significantly reduce or eliminate weaknesses identified in the accounting cycles biometric tools were recommended.

Additionally, the biometric tools were recommended, such as fingerprints, voice or facial recognition were analyzed to avoid conflicts with individuals' rights. The recommended biometric tools revealed possible violations of several constitutional amendments. For instance, the First, Fifth and Ninth Amendments could possibly be violated by implementing certain biometrics. However, additional recommendations were given to avoid the violation of the Bill of Rights. For example, one of the recommendations to remedy the problem was to gain the consent of the employees with regard to introducing biometrics in the workplace.

Throughput model

The Throughput Model is composed of four components: information, perception, judgment and decision (Rodgers et al., 2009) (Figure 6.3), whereby, perception is how an individual frames the circumstances. Information is the available facts that will influence the decision making process. Judgment is the process of comparing and weighing different alternatives. Lastly, is the decision made or not made by the individual. In addition, the Throughput Model main ethical algorithmic pathways are: ethical egoism, deontology, utilitarianism, relativist, virtue ethics and ethics of care (Rodgers, 2009).

Figure 6.3 illustrates the different ethical pathways to reach a decision.

1. P → D Ethical Egoism
2. P → J → D Deontology
3. I → J → D Utilitarianism
4. I → P → D Relativist
5. P → I → J → D Virtue Ethics
6. I → P → J → D Ethics of care

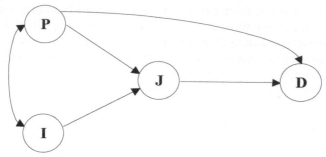

Figure 6.3. Throughput Modelling diagram. Where P = perception, I = information, J = judgment, and D = decision choice.

After all of these factors (i.e., fraud issues, biometric tools, and Bill of Rights issues) were analyzed, the accounting cycles were categorized accordingly by using the Throughput Model's ethical pathways. Overall, the current accounting cycles were categorized as deontology (P→J→D). However, with the recommendations (additional information) proposed, the cycles were reclassified as ethics of care (I→P→J→D). Also, the introduction of biometrics to enhance internal controls demonstrated a utilitarian approach (I→J→D). That is, to better protect the employer and employee. Lastly, a conclusion was reached, reinforcing the importance of internal controls and how biometrics can assist Advanced Intelligence's control environment.

Company profile of advanced intelligence stores

Advanced Intelligence became a public trade company in 1980. Advanced Intelligence must comply with laws and regulations subject to companies that trade in the US Stock Exchange. An important requirement that public companies must meet is the adoption and maintenance of adequate internal controls. This requirement was a highlight in the Sarbanes-Oxley Act of 2002 that was enacted due to the several financial scandals (i.e., Enron, WorldCom, Arthur Andersen, etc.), that occurred in the early 20th century. The Sarbanes-Oxley Act requires the signatures of key company executives (CFO, CEO, etc.), to assure the adoption and maintenance on the company's internal controls (Securities and Exchange Commission, 2012). Therefore, an analysis was conducted on Advanced Intelligence's current internal controls on fundamental areas of the company, such as inventory, purchasing/cash disbursements and sales/cash receipts. Additionally, weaknesses in the internal control systems were identified and attempted to be solved by implementing biometric tools. Further, the biometric tool recommendations were analyzed to avoid potential conflicts with the Bill of Rights. Lastly, the accounting cycles were classified under the Throughput Model. Figure 6.4 demonstrates the course of the analysis of Advanced Intelligence.

The first accounting cycle that requires examination is the cash receipts. Generally speaking, the sales/cash receipts is one of the most

Sales and Cash Receipts

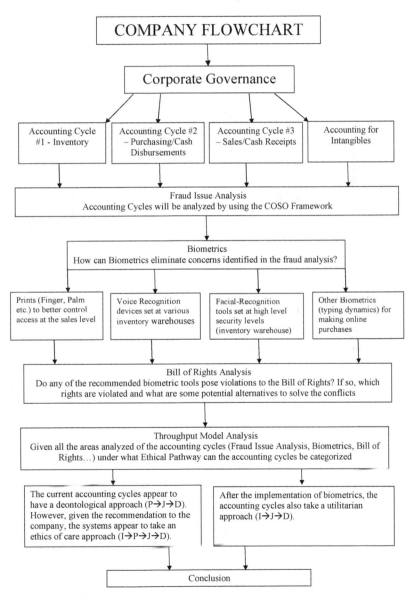

Figure 6.4. Accounting cycles and internal controls

important organizational assets. Typically, sale clerks handle a very high level of sales throughout in most of their shifts. As a result, it is common to have shortages/overages problems at the end of the day. Further, on a typical day, shortages and overages are insignificant in terms of amounts. For example, in a given week the net total shortage/overage is less than twenty dollars. Clerks log in to cash registers by entering their assigned passwords. One note of concern is that lately there have been issues regarding employees using other employees' log in information.

Purchases and cash disbursements

Further, another accounting cycle that requires examination is the cash disbursements. Overall, the major concern in the purchase/cash disbursement cycle is the proper authorization to pay vendors. As explained previously, Advanced Intelligence operates various plants nationwide and internationally. Hence, Advanced Intelligence deals with various vendors to pay operating expenses, such as utility costs. Another significant concern is the purchasing of inventory. That is, the company wants to ensure that only the necessary inventory is purchased from legitimate vendors. In the industry that Advanced Intelligence operates, it is very common to have bogus vendors' fraud-related issues. In addition, the majority of purchases and payments are completed online for convenience. This is different from check payments, which can be lost in the process or can arrive late and consequently lead to unsatisfied vendors. Currently, only certain designated accounts payable clerks in the purchasing/cash disbursements departments have the authorization to make purchases/payments. Accounts payable clerks accomplish this task by logging (entering assigned passwords) into the accounts payable system.

Inventory

Lastly, the inventory accounting cycle requires examination. In respect to inventory the major concern is to protect the inventory from theft. Due to the high level of inventory the business handles, it is difficult to secure entrances to inventory warehouses. Currently employees must enter an assigned password into devices placed at the warehouse entrances in order to gain access. However, employees have developed a bad habit; that is employees sometimes leave the door to the warehouse open. Employees claim it is not practical to enter their passwords every time they need to gain access to the warehouse because inventory personnel have to meet strict deadlines, such as loading/unloading inventory.

Intangibles

Advanced Intelligence's intangibles mainly consist of goodwill. Moreover, Advanced Intelligence does not amortize goodwill. Instead, the accounting department evaluates goodwill annually and adjusts the fair value based on its evaluation method (Wikinvest, 2008). Further, Advanced Intelligence carries goodwill at fair value cost. Intangibles must be analyzed carefully, since if not treated properly, these assets can create material misstatements. Finally, unintentionally, intangibles can be treated improperly, because of the nature of intangibles. That is these types of assets can sometimes be difficult to measure in monetary terms.

Fraud analysis

The fraud triangle identifies three components that cause an individual to commit fraud (Association of Certified Fraud Examiners, 2013; Rodgers, 2012). The three components are pressure, rationalization and opportunity (ACFE, 2013). Pressure represents the motivation for an individual to engage in fraudulent behavior. For instance, an employee that has a gambling addiction could meet the characteristics for pressure. That is, the employee may be motivated to steal money from the company to satisfy his or her gambling expenditures. Rationalization (or attitudes/personality features) deals with the individual justifying the fraudulent behavior (i.e., stealing money from the company). According to the Association of Certified Fraud Examiners many individuals justify their crimes by saying things to themselves, such as "I was entitled to the money, I was only borrowing the money, or I was underpaid; my employer cheated me" (2013). Lastly, opportunity consists of the available methods that the fraudster can use to commit fraud. Opportunity is closely related to the employee's position in the company. For instance, a manager has a higher opportunity to commit fraud than a normal employee, because the manager can override internal control procedures and easily access multiple assets.

Moreover, a possible fraud issue in the inventory accounting cycle is the misappropriation of assets. Employees with access to the warehouse have the opportunity to steal company assets (inventory). Also, other individuals have the opportunity to steal inventory due to the inventory personnels' bad habits. As mentioned earlier, employees tend to leave entrances to the warehouse unsecure. According to the ACFE, misappropriation of assets is among the most common type of fraud (Hall, 2011). Similarly, in the sales/cash receipts internal control system a potential fraud issue is the misappropriation of assets. However, in this particular case the asset susceptible to theft is cash at hand from cash registers. In both internal control systems, the opportunity is present. The components of the fraud triangle are interrelated; however, opportunity is the most powerful component. An employee can have all kinds of pressures and rationalizations to commit fraud, but if the opportunity (the means) is not present, then the fraud cannot occur. In addition, in the purchasing of internal control system, one weakness is the authorization of purchases. Since, the system only requires a password to log in into the purchasing system, any user (authorize/unauthorized) can make purchases.

Additionally, another area that requires a fraud analysis is management. Since, Advanced Intelligence is a public company; there might be an incentive to manipulate financial statements to meet desirable expectations. For instance, management is given a bonus based on the performance (profits) of the company. Hence, management might try to artificially increase profits, assets or decrease liabilities. Management fraud is sometimes hard to identify given that management can override internal control procedures. Nevertheless, the triangle fraud can assist in identifying potential fraudsters. Management will have the opportunity to commit the fraud due to their position of power. However, the other two components of the fraud triangle (rationalization and pressure) can be analyzed to management personnel. This is where the typical questions come in to use to possibly detect management fraud. Some of the typical questions are the following:

- Do management personnel have a gambling history?
- Do management personnel seem to be living beyond their means?
- What type of work ethic does management practice?
- What is the tone at the top?

Another method that can be very helpful to identify management fraud is to study financial statements periodically. This method can assist in revealing unusual or rapid increases/decreases in key financial statement items, such as profits, assets or liabilities.

Biometrics

Biometrics is the measurement of an individuals' characteristic, such as fingerprints, for identification purposes (Rosenzweig et al., 2004). Moreover, the fraud issues identified in Advanced Intelligence's internal control systems can be solved by the implementation of various biometric tools. In respect to the purchasing/cash disbursement, Advanced Intelligence can strengthen their system by introducing typing dynamics and fingerprint scanning. That is, a major weakness in Advanced Intelligence's purchasing system is purchase authorization. Typing dynamics can be implemented to assure that only authorize employees make the require purchases via online. More importantly, typing dynamics can overcome the current weakness, which is an unauthorized user can obtain the password and gain access to the purchasing system. Typing dynamics will be more difficult for fraudsters to replicate. Also, another biometric tool that will enhance the internal controls of the cash disbursements cycle is for managers to sign-off (using fingerprints) on the work performed by Accounts payable clerks. Figure 6.5 illustrates how the biometric tool (typing dynamics and fingerprints) will be implemented into the purchasing/cash disbursement cycle.

In addition, inventory can be better protected by introducing a facial recognition device at the various inventory warehouses. Besides improving internal controls for inventory, facial recognition will not delay business procedures. As noted earlier, employees claim that it is too time consuming to enter passwords each time they need to enter the warehouses. In contrast, the facial recognition device will identify individuals as they are working. In other words, the camera that will be installed for facial recognition will identify employees as they enter the

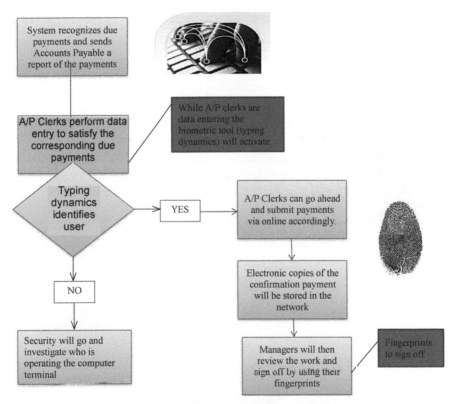

Figure 6.5. Purchasing/Cash disbursements. This figure illustrating the cash disbursements cycle with the introduction of biometrics

warehouse. More importantly, employees will not have to stop to enter passwords. Figure 6.6 illustrates how the biometric tool (facial recognition) will be implemented into the inventory cycle. Similarly, another biometric tool that will have similar results is voice recognition.

The biometric tool that can improve the internal controls of the sales/cash receipts cycle is fingerprints. As identified in the cash receipts fraud analysis, a major weakness is that multiple employees can operate multiple cash registers. Instead of a simple password, the cash registers can have fingerprint scanning devices to log into. The biometrical tool can improve control on who is operating the various cash registers. Ultimately fingerprints will make employees more accountable for their cash registers. Figure 6.8 illustrates how the biometric tool (fingerprints scanning) can be implemented into the sales/cash receipts cycle.

Moreover, it is important to note and explain the categories of the suggested biometric tools. That is, biometrics can either be categorized as physiological or behavioral. Physiological biometrics means that the

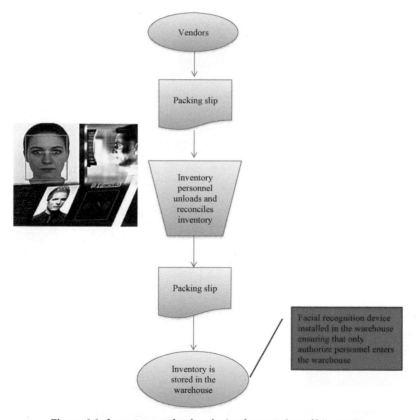

Figure 6.6. Inventory cycle after the implementation of biometrics

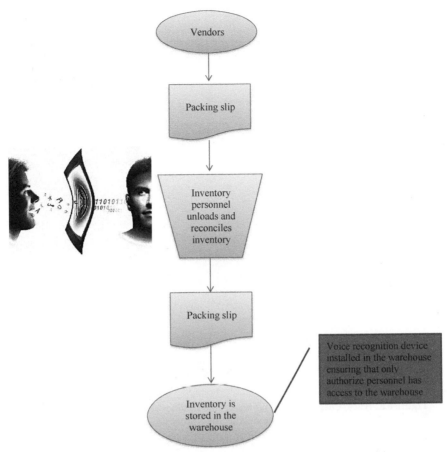

Figure 6.7. This illustrates how the biometric tool (voice recognition) will be implemented into the inventory cycle

individuals' characteristics measured will not change significantly over time (Rosenzweig et al., 2004). An example of a physiological biometric is fingerprints. An individuals' fingerprints will remain same throughout his life time, unless something extraordinary happens (i.e., cuts or burns on fingerprints). On the other hand, behavioral biometrics will change significantly over the lifetime of an individual (Rosenzweig et al., 2004). An, example of this category is voice recognition. An individuals' voice will be very different from adolescence to adulthood. Depending on the category the biometric tools can represent advantages or disadvantages. For example, one of the recommendations is to instal voice recognition devices (behavioral biometrics) at the multiple warehouses. In the event that an authorized employee has cold/flu symptoms (sore throat) their

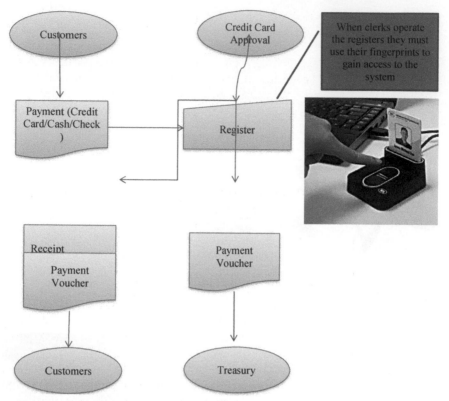

Figure 6.8. Sales/Cash receipts. This figure illustrating the cash receipts cycle with the implementation of biometrics.

voice will be affected. Hence, the individual will not be granted access to the warehouse and daily business procedures will be disrupted.

Privacy issues

As a result of technological advances such as biometrics, the sensitive topic of employees' privacy and rights has created controversy in the workplace. In other words, enhancing internal control systems could potentially violate amendments in the constitution. Some of the recommendations to strengthen Advanced Intelligence's internal control systems are to implement typing dynamics, fingerprint scanning and voice recognition. Of the various biometrics suggested, voice recognition could pose the biggest threat to employees' privacy and rights (Rodgers, 2010, 2012). For example, if an employee gets fired for saying something confidential around the voice recognition system. Depending on the circumstances of the situation, this could potentially violate the right to freedom of speech.

A way to avoid this dilemma in the workplace is to inform employees and make them sign a contract, specifying that anything that is captured in the voice recognition system can be used to reprimand or fire employees. Another potential issue is fingerprint biometric tools could affect the The United States Fifth Amendment. Assume that the employer (Advanced Intelligence) inappropriately uses an employee's saved fingerprints to commit a crime. The employee could face charges and go to jail. However in most cases, due to the Fifth Amendment (i.e., self-incrimination) the authorities cannot use biometrics (fingerprints) to apprehend the employee.

Further, the Fourth Amendment (i.e., no unreasonable searches and seizures) could potentially be violated as well. For instance, again assume that the voice recognition device captures an employee saying something "suspicious." The employer might issue a search against the particular employee. In addition, any biometric tool (i.e., fingerprints, facial recognition, etc.), that obtains more information than the intended information might be a violation of the Fourth Amendment. This scenario can also lead to violation of the Third Amendment (i.e., no quartering of troops). In this context, the employer cannot obtain employees' information/personal property without their consent. Finally, the *Ninth Amendment* (i.e., just because it is not in the constitution does not mean you have it) is another right that might be violated when employing certain biometrics. Due to the nature of biometrics, it requires personal information/property from the individual. This may indicate a situation affecting the Ninth Amendment. As explained earlier, the best way to avoid these types of dilemmas is to gain the consent of the employees in terms of implementing biometrics in the workplace. Also, the employer must ensure that the personnel in charge of biometric tools use the equipment appropriately.

Overall, the current internal control systems of Advanced Intelligence can be categorized under deontology. That is, the internal control systems are mainly duty- and rule-based. However, given the recommendations to strengthen Advanced Intelligence's internal control systems, systems lean to take an ethics of care approach. The ethics of care pathway is built on deontology, except that ethics of care is guided through additional information. This represents the recommendations (additional information) given to Advanced Intelligence to enhance the internal control systems. The ethics of care pathway is similar to the stakeholder theory. The stakeholder theory suggests that an organization/company will take any measures necessary to avoid harming stakeholders (i.e., employees, community, lenders, vendors, creditors, stockholders, etc.). Since, Advanced Intelligence is a public company, the ethics of care pathway fits perfectly with the business structure. As a public company, auditors of Advanced Intelligence must place the interests of the public first.

Also, the introduction of the various biometric methods the internal control systems could take a utilitarian approach. The implementation of biometric techniques to better identify and control authorized users creates the greatest good for the greatest number of individuals. For example, if only passwords were required to log in to cash registers or purchasing systems, an employee could easily obtain the passwords and engage in fraudulent activities. Additionally, innocent employees would get accused, because the illegal activities were committed under their user name and password. Whereas in biometrics, such as fingerprints, voice or facial recognition was used as passwords, the fraudsters would have a difficult time obtaining access to the company's system. Consequently, the company's assets are better protected by using biometrics than simple passwords. Thus, the greatest good (employees and assets are better protected) and for the greatest number of people (employees and employer) is created.

To sum up, in the digital age, security is one of the primary concerns of any organization. Every organization has realized that data is a major resource. Hence, organizations deploying advanced security mechanisms such as biometrics to safeguard business data will be the future winners for efficiency and effectiveness of operations.

Conclusion

In tackling fraud, organizations must begin by discovering and assessing their baseline risks and investigating the different types of fraud to which they are vulnerable. While external threats such as cyber criminals may attract the most attention, organizations cannot ignore risks closer to home.

Prevailing over internal fraud is a strategic challenge. Organizations' managers and executives must employ anti-fraud programs which encompass training for all employees, especially those in high-risk areas. This can assist personnel to recognize dubious activities and areas that malicious actors could possibly manipulate. Moreover, it provides an organization to communicate its commitment to high ethical standards and procedures as well as fraud prevention.

Organizations can make it more challenging to commit fraud by modifying their internal environmental factors; focusing on techniques to shrink the opportunity to commit these fraudulent acts. This is where an organization would benefit from visiting their formal/informal counter fraud measures and amending them accordingly.

In sum, those who commit so-called 'crimes of opportunity', factors such as corporate culture, morale, reduction of temptation and reduction of opportunity play a momentous role.

Machine learning has the capability for 24/7 monitoring and handles larger data loads than any individual can deal with in an organization. Machine learning greatly reduces the workload of security analysts and cyber security personnel. What machine learning is especially good at is identifying and hence filtering out potential threats that expose known patterns, in large quantities. If a subject expert undertook and analyzed these patterns, s/he would find them highly complex, involving a large amount of repetitive work. With a sufficient number of samples, machine learning can replicate the decision-making process of a security expert, thereby afterwards be deployed at scale to classify new samples.

The Advanced Intelligence case indicated that with growth, technological advances, laws and regulations, internal control systems has become more fool proof against fraudulent activities. Specifically, the fraud analysis performed on the purchasing/cash disbursements, inventory and sales/cash receipts internal control systems revealed the potential weaknesses. Given the fraud issues identified in the fraud analysis, biometric tools were recommended to significantly reduce or eliminate fraud issues. However, the recommendations of biometrics required a closer examination of employees' rights and privacy (Bill of Rights). Once the biometric methods were analyzed and modified to eliminate possible violation of employees' rights or privacy the internal control systems were categorized under the ethical pathways of the Throughput Model. As evident from the case presented in this chapter, the Throughput Model can assist Advanced Intelligence's management in carrying out the introduction of biometrics along with algorithmic pathways in the workplace.

In summary, biometrics is smart; however, biometrics enhanced with Artificial Intelligence tools increases the efficiency of the control system. Biometrics such as fingerprint, facial recognition and iris scans are being implemented for authenticating employees at the workplace and identifying smartphone owners. Such biometrics can be used in organizations to authorize data access for confidential data.

References

ACFE. 2013. The Fraud Triangle. Retrieved April 3, 2013 from http://www.acfe.com/fraud-triangle.aspx.

Hall, J.A. 2011. Accounting Information Systems (8th edn.). Mason OH: South Western.

Iansiti, M. and Lakhani, K.R. 2017. The Truth about Blockchain. Harvard Business Review (January-February): 118–127.

Kochems, A., Rosenweig, P. and Schwartz, A. 2004. Biometric Technologies: Security, Legal, and Policy Implications. Retrieved April 3, 2013 from http://www.heritage. org/research/reports/2004/06/biometric-technologies-security-legal-and-policy-implications.

Prathyusha, T., Kavya, M. and Akshita, P.S.L. 2018. International Journal of Computer & Mathematical Sciences, 7(3): 232–237.

Rodgers, W. 2009. Ethical Beginnings: Preferences, Rules, and Principles Influencing Decision Making. NY: iUniverse, Inc.

Rodgers, W., Guiral, A. and Gonzalo, J.A. 2009. Different pathways that suggests whether auditors' going concern opinions are ethically based. Journal of Business Ethics, 86: 347–361.

Rodgers, W. 2010. E-commerce and Biometric Issues Addressed in a Throughput Model. Hauppauge, NY: Nova Publication.

Rodgers, W. 2012. Biometric and Auditing Issues Addressed in a Throughput Model. Charlotte, NC: Information Age Publishing Inc.

Rodgers, W., Söderbom, A. and Guiral, A. 2014. Corporate social responsibility enhanced control systems reducing the likelihood of fraud. Journal of Business Ethics, 91(Supplement 1): 151–166.

Royal Academy of Engineering. 2017. Algorithms in decision-making A response to the House of Commons Science and Technology Committee inquiry into the use of algorithms in decision-making (April). https://www.raeng.org.uk/publications/responses/algorithms-in-decision-making.

Securities and Exchange Commission. 2012. The Laws that Govern the Securities Industry. Retrieved May 7, 2013 from http://www.sec.gov/about/laws.shtml.

Wikinvest. 2008. Goodwill and Other Acquired Intangible Assets. Retrieved May 7, 2013 from http://www.wikinvest.com/stock/Advanced Intelligence_(WMT)/Goodwill_Other_Acquired_Intangible_Assets.

7

Artificial Intelligence, Biometrics, and Ethics Examples

"Artificial Intelligence at maturity is like a gear system with three interlocking wheels: data processing, machine learning and business action. It operates in an automated mode without any human intervention. Data is created, transformed and moved without data engineers. Business actions or decisions are implemented without any operators or agents. The system learns continuously from the accumulating data and business actions and outcomes get better and better with time."

—Niranjan Krishnan, Head of Data Science at Tiger Analytics

Traditional security and authentication methods have been created to put up security measures for individuals, challenging them at several points throughout in order for them to identify themselves. To do this they must enter, remember, and frequently change passwords, and in the expanding world of two- or multi-factor authentication, an organization seeks enhance security, which could reduce productivity. Nevertheless, traditional passwords and other authentication methods can be compromised, so the impact to user processes does not always deliver proportional security.

Artificial Intelligence and biometrics can transform how authentication takes place. This is done by supporting enriched security via automatically identifying employee identity without impacting workflows and user processes.

As Artificial Intelligence becomes more omnipresent, so too has the concern over how we can have confidence that it reflects ethical values. An example that gets cited repeatedly to demonstrate how challenging this can be is the moral or ethical decision choice an autonomous automobile might have to make to prevent a collision. What if there is a truck coming towards a driver who has to turn sharply to steer clear from being hit and critically injured; however, the automobile will hit a baby if it swings over to the left and an elderly person if it swerves right. Hence, what should the autonomous car do?

In other words, without the appropriate care in programming Artificial Intelligence technology, a system may possibly have the bias or predisposition of the programmer, which can play a part in determining outcomes. Further, machines may become biased due to the training data they are fed, which may not be fully representative of what one is attempting to teach them.

Ethical systems or positions can help in developing frameworks for thinking about these types of issues. Some issues involve an ethical dilemma, which is a very complicated topic, one that should incorporate cooperation with other technology organizations. To further complicate efforts to instill ethical or morality in Artificial Intelligence systems, there is no universally accepted ethical system for Artificial Intelligence. Therefore, this chapter discusses six dominant ethical positions used globally by individuals and organizations (Rodgers, 2009; Rodgers and Al Fayi, 2019). This chapter can provide assistance to the Artificial Intelligence community by asserting six dominant ethical positions that can address the transparency issue. To trust computer decisions, ethical or otherwise, individuals need to know what type of ethical position an Artificial Intelligence system arrives at in its conclusions and recommendations. Right now, deep learning does poorly in this regard. Nonetheless, Artificial Intelligence systems can address this issue by employing ethical algorithmic pathways that deals with text documents that are part of knowledge bases from which Artificial Intelligence systems draw their conclusions.

Biometric recognition is an integral part of our living. This chapter deals with machine learning methods for recognition of individuals based on fingerprints and face biometrics. The main purpose of machine learning, neural networks, and deep learning topics are to reach a state when machines (computers) are able to respond without individuals' explicitly programming them. These areas are closely related to Artificial Intelligence, knowledge discovery, data mining, big data and neurocomputing. The case presented in this chapter includes relevant machine learning methods with the main focus on neural networks and deep learning applied to biometrics. Some aspects of theory of neural networks are addressed such as visualization of processes in neural networks and deep learning, internal representations of input data as a basis for feature extraction methods and their applications to image compression and classification for the auditing case. Machine learning methods can be efficiently implemented for feature extraction and classification and therefore are directly applicable to biometric systems. Biometrics deals with the recognition of people based on physiological and behavioral characteristics. Biometric recognition uses automated methods for recognition, and this is why it is closely related to machine learning.

Fingerprints technology is discussed in the case presented in this chapter. A fingerprint scanner system has two basic jobs. First, it needs to get an image of the individual's fingers. Second, it needs to establish whether the pattern of ridges and valleys in this image matches the pattern of ridges and valleys in pre-scanned images.

Only specific characteristics, which are distinctive to every fingerprint, are filtered and saved as an encrypted biometric key or mathematical representation. No image of a fingerprint is ever saved, only a series of numbers (a binary code), which is implemented for verification. The algorithm cannot be reconverted to an image; hence, no one can replicate a person's fingerprints.

Face recognition is discussed in this presentation in that it covers the aspects of face detection, detection of facial features, classification in face recognition systems, state-of-the-art techniques in biometric face recognition, face recognition in controlled and uncontrolled conditions and single-sample problem in face recognition.

This chapter is divided into the following sections: (1) ethical process thinking method, (2) matters pertaining to Artificial Intelligence impact on privacy and ethical considerations, (3) rule-based and utilitarian-based algorithmic pathways, and discussion of (4) automobile parts company using ethical platforms and biometric tools.

Ethical Process Thinking Approach

As discussed in Chapter 6, the Throughput Model (Figure 7.1) can display an *ethical process thinking approach* (Rodgers, 2009), for an Artificial Intelligence biometric system. This modelling process begins with the individuals stating their views of what should be done. The advantage of this approach is that it helps users and implementors of algorithms to understand why individuals can be influenced by algorithms that supports their position, and have ignored other information which does not support their position. This approach helps uncover the observations and values that individuals rely upon when taking positions on issues. Also, the model is useful in depicting latter stages of processes, such as judgment, that are implemented in supporting individuals' positions.

Based on Figure 7.1, we can establish six algorithmic ethical pathways (Rodgers, 2009):

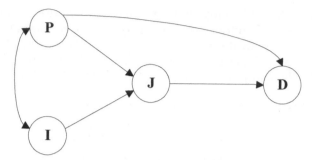

Figure 7.1. Throughput Modeling: Ethical process thinking model (Rodgers, 2009). Where, P = perception, I = information, J = judgment and D = decision choice

(1) **P → D** *ethical egoism (preference-based) position*
(2) **P → J → D** *deontology (rule-based) position*
(3) **I → J → D** *utilitarian (principle-based) position*
(4) **I → P → D** *relativist position*
(5) **P → I → J → D** *virtue ethics position*
(6) **I → P → J → D** *ethics of care (stakeholder-based) position*

Concerns Regarding Artificial Intelligence Impact on Privacy and Ethical Considerations

Many concerns have grown over how Artificial Intelligence may impact privacy, cyber security, employment, inequality, and the environment. Customers, employees, boards, regulators, and corporate partners are all querying the same question: Can we trust Artificial Intelligence? Hence, it is no wonder that executives say ensuring that Artificial Intelligence systems are trustworthy is top priority. How they will overcome that challenge rests on whether they can address the following facets of responsible Artificial Intelligence:

1. *Fairness*: Are we minimizing bias in our data and Artificial Intelligence models? Are we addressing bias when we use Artificial Intelligence?

2. *Interpretability*: Can we explain how an Artificial Intelligence model makes decisions? Can we ensure that those decisions are accurate?

3. *Robustness and security*: Can we rely on an Artificial Intelligence system's performance? Are our Artificial Intelligence systems vulnerable to cyber attack?

4. *Governance*: Who is accountable for Artificial Intelligence systems? Do we have the proper controls in place?

5. *System ethics*: Do our Artificial Intelligence systems comply with regulations? How will they impact our employees and customers?

The discipline of Artificial Intelligence ethics is likely to differ basically from the ethical discipline of noncognitive technologies, in that:

(a) The local, precise nature of the Artificial Intelligence may not be predictable apart from its safety, even if the programmers do everything right;

(b) Attesting to the safety of the system becomes a greater challenge since we must verify what the system is trying to perform, rather than being able to prove the system's safe behavior in all operating contexts;

(c) Ethical cognition ought to be procured as a subject matter of engineering.

Transparency issues in Artificial Intelligence

Transparency is a necessary feature of Artificial Intelligence. Moreover, it is also essential that Artificial Intelligence algorithms taking over social functions be predictable to those they govern. That is, when Artificial Intelligence algorithms take on cognitive work with social dimensions, then the Artificial Intelligence algorithm inherits the social requirements cognitive tasks heretofore performed by individuals.

Artificial Intelligence algorithms should be robust against manipulation. For example, a warehouse biometric facial recognition system to scan the warehouse for unwanted personnel must be robust against Type 2 errors, which indicates a person deliberately wearing a face mask in order to exploit flaws in the algorithm. That is, the face mask could be a facsimile of a person authorized to be in the warehouse. Robustness against manipulation is a common criterion in information security; nearly the criterion. A Type 2 error (i.e., false accept rate or false match rate) occurs when a biometric system wrongly authenticates the identity of a person and grants them access to what the system is safeguarding. On the other hand, Type I error implies the non-acceptance of a biometric feature that ought to be accepted.

Artificial Intelligence systems have been operating in various environments. Its employment in many transactions across the internet is based on consent. The existing legal framework does not deter public and private actors from putting into practice applications. The installment of an Artificial Intelligence system does not threaten procedural rights (i.e., rights in a court of law); their use is deemed intrusive but within reasonable limits. Nonetheless, their widespread implementation and the fear of an "informational biometric" may have a psychological effect. The following four premises are illustrated in order to afford an enhanced understanding of the legal implications of an Artificial Intelligence system:

1. *Facilitating legal environment*: The existing legal environment (privacy and data protection) is accommodating in that it permits legislation legitimizing the *de facto* business use of personal data. Data protection rules regulate the utilization of Artificial Intelligence systems; nonetheless, they require more ethical debate.

   For machine learning to distinguish noteworthy patterns in the present and predict the future, it must be communicated effectively to individuals. For example, displaying a sufficient amount of historical data to an Artificial Intelligence system on consumer behavior. Therefore, it will eventually be able to predict how consumers and others who are comparable will behave going forward. Nonetheless, to create the data sets required for training, the data must be labeled. Another example is defining whether a consumer is satisfied with the product or service. For those data sets to assist supporting Artificial Intelligence across an organization (i.e., those consumers may interact with more than one business line), standards will be compulsory for labeling them consistently.

   Artificial Intelligence ethical standards can generate and monitor data standards, as well as improve systems and processes that affords better usage for employees as well as labeling data sets for future use.

2. *Opacity/transparency rules compulsory enforced*: Data protection (transparency rules) does not explain the limits of utilization and exploitation of Artificial Intelligence systems. That is, opacity (privacy) rules may retard use in cases where there is the requirement to certify against outside disproportionate power balances. Other ways to formulate Artificial Intelligence more trustworthy can emanate from upgradation in Artificial Intelligence.

3. *Widespread Artificial Intelligence systems into operation fosters basic concerns*: As Artificial Intelligence systems are disseminated in the global community some apprehensions are of tremendous concern. For example, alarms pertaining to (a) power enlargement, (b) continuing implementation of existing data, (c) identifiable threats related to the use of Artificial Intelligence systems by the public sector, and (d) failure to protect individuals from their penchant to trade their own privacy with what appears to be an exceedingly inexpensive convenience.

4. *Engagement of Artificial Intelligence in law enforcement*: It is imperative that Artificial Intelligence type-evidence be regulated when presented as evidence in courts of law so as to adequately safeguard suspects (e.g., being listened to, right to counter-expertise). Developing regulations encircling data privacy will affect Artificial Intelligence. Also, how it may constrain its evolution since it shapes how organizations

operating globally can utilize data generated across regions. Europe's General Data Protection Regulation (GDPR) became law in May 2018, and the California Consumer Privacy Act (CCPA) may become a law in 2020. Although, GDPR and CCPA may have differences, they both offer individuals the right to see and control how organizations collect and implement their personal data. Moreover, there is legal recourse if they experience damages due to bias or cyber security breaches.

Rule-based and Utilitarian-based Algorithmic Pathways

Rule-based (P→J→D) algorithmic systems

According to Rodgers and Gago (2001), deontology (i.e., rule-based) emphasizes the rights of individuals and on the judgments associated with a particular decision process rather than on its choices. Moreover, the 'deontology' viewpoint (P→J→D) emphasizes individuals' rights as "a basic premise to this viewpoint is that equal respect must be given to all individuals". According to Rodgers (2011), "Deontologists also regard the nature of moral principles as permanent and stable, and that compliance with these principles defines ethicalness." People who make decisions based on this model "regard the nature of moral principles as permanent and stable, and that compliance with these principles defines ethicalness. Furthermore, they believe that individuals have certain rights, which include (1) freedom of conscience, (2) freedom of consent, (3) freedom of privacy, (4) freedom of speech, and (5) due process" (Rodgers, 2011). Therefore, an organization ought to spell out the rules underlining an algorithmic system that supports its Artificial Intelligence application.

Nonetheless, one of the most important functions of an algorithmic rule-based system is to be predictable, so that, e.g., contracts can be written knowing how they will be executed. One of the main purposes of a rule-based system is not necessarily to optimize a particular function, but to provide a predictable environment within whereby individuals can effectively improve their performance.

Utilitarian-based (I→J→D) algorithmic systems

According to the article "Cultural and Ethical Effects on Managerial Decisions" by Rodgers and Gago (2001), individuals tend to follow different pathways when embracing an ethical decision. One of the most common ethical pathways is the utilitarian perspective. Under the utilitarian-based doctrine, an ethical decision is considered as one which brings the greatest good to the greatest number of persons. Further, it can be interpreted as the decision choice whose outcome causes the least amount of harm to society. This kind of ethical reasoning is traditionally related to politics, where societal leaders are forced to make decision choices affecting a great number of people. According to Rodgers and Gago (2001), a utilitarian pathway approach involves the processing of information without bias (perception) followed by judgment. The individual's capacity to analyze information, and not his or her perception, concludes what the best outcome could be in any given situation. Thus, the individual does not frame a scenario based on previous experience, knowledge or training but rather on purely circumstantial information. The quality of the decision is directly dependent on the amount of information available, time to collect and analyze information, the quality of the information and its relevance to the decision maker. In essence, a simplified utilitarian model can be compared to a cost versus benefit analysis, which is a traditional form of decision-making in the global community.

The prospect of Artificial Intelligence and machine learning techniques is that it can enhance core efficiency and effectiveness that pertains to business and non-business activities of the global community (The National Artificial Intelligence R&D Strategic Plan, 2016). Nonetheless, to procure the societal benefits of Artificial Intelligence systems, complying with an ethical framework is paramount. Practical and research efforts ought to be concentrated on implementing regulations in Artificial Intelligence system design that are updated on a continual basis to respond appropriately to different application fields and actual situations. Further, a safe and secure Artificial Intelligence system is one that acts in a controlled and well-understood manner. The design philosophy must be such that it ensures security against external attacks, anomalies and cyberattacks.

Artificial Intelligence with an ethical emphasis reflects the creation of intelligent machines that work and react like humans, built with the ability to autonomously conduct, support or manage business activity across disciplines in a responsible and accountable way. At its core, Artificial Intelligence is the creation of intelligent machines that think, work and learn like humans. Artificial Intelligence should not be a replacement for standard ethical values, principles, rules or procedures.

Individuals possess inherent social, economic and cultural biases. Unfortunately, these biases permeate the social fabrics around the world. Therefore, Artificial Intelligence offers a chance for the business community to eliminate such biases from their global operations.

Ethics and law are inextricably linked in our global society, and many legal decisions arise from the interpretation of many ethical issues. Artificial Intelligence adds a new dimension to these questions. Systems that implements Artificial Intelligence apparatuses are becoming increasingly more autonomous in terms of the complexity of the chores they can perform, their potential impact on global society and the shrinking capability of people to understand, predict and control their functioning.

Most individuals underestimate the valid level of automation of these systems, which have the ability to learn from their own experience and perform actions beyond the scope of those intended by their inventors.

Discussing the ethics of Artificial Intelligence, though, is far more complex than simply agreeing, for instance, not to fly your drone into a person. This instigates a variety of ethical and legal difficulties that will be encapsulated in this chapter as six dominant ethical algorithmic pathways.

The onus is on the Artificial Intelligence community to build technology that employs data from relevant, trusted sources to embrace a diversity of culture, knowledge, opinions, skills and interactions. Artificial Intelligence operating in the global community today performs repetitive tasks well, learns on the job and even incorporates human social norms into its work. Nonetheless, Artificial Intelligence also consumes a substantial amount of time combing the web and its own conversational history for more context that will inform future exchanges with human counterparts.

This pervasiveness of data sets and incomplete information on the internet presents a challenge and an opportunity for Artificial Intelligence developers. When built with responsible organizational and social practices in mind, Artificial Intelligence technology has the potential to ethically deliver products and services to individuals who need them, thereby reducing the omnipresent human threat of bias.

Moreover, policy initiatives should explicitly touch upon building an incubatory environment for Artificial Intelligence-based research and training. This includes making effective training data sets from various portals available to academicians and the public at large. In due course, the complexity and sophistication of Artificial Intelligence systems in delivering outputs will be a factor of the quality of training data fed into the system. Open software libraries, toolkits and development tools with low-cost code repositories and development languages such as "R" open source programming language and Python are decreasing the barriers to the use and extension of Artificial Intelligence systems.

Example: Automobile Parts Company Incorporated (APCI)

This section explores a company called the Automobile Parts Company Inc. (APCI), which utilizes biometrics technology in their organization. Biometric technology is suggested to be implemented in different areas in order to improve its existing internal control system and to prevent fraud. Utilization of biometrics is recommended in the purchasing department, inventory control department, payroll department and access control dept. In order to protect intangible or knowledge assets (Rodgers, 2003, 2007; Tillquist and Rodgers, 2005). Biometrics technology enables APCI to maintain a superior market position, which was attained from its continuous efforts in innovation through research and development

(R&D). In part, this is accomplished by safeguarding its trade secrets as well as reducing fraud in operations by way of decreasing the "opportunity" factor in the "fraud triangle" as discussed in Chapter 6 (Rodgers et al., 2014). Despite biometrics tools great advantage, ethical issues should be considered due to the biometrics challenges prior to its implementation. In this section, ethical issues are analyzed from two different perspectives: utilitarianism and deontology. While biometrics provide sundry benefits such as prevention and detection of fraud, improved cost accounting, strengthened accountability among employees, and protection of trade secrets, it may violate employees' privacy. Potential invasion of privacy from the utilization of biometrics at workplace can be resolved by establishing policies and procedures on the proper use of biometrics (see Figure 7.2).

APCI is well positioned in terms of competitive costing and strategic positioning in the global market. Its vision is to move forward through innovation in technology for customer satisfaction in order to enhance its future. With consistent support from the Research and Development (R&D) department, APCI has obtained many patents for its unique and superior automotive parts as well as trade secrets that are stored in the database.

Use of biometrics in internal control

APCI's Internal control system is in place in order to prevent and detect fraud. The flow chart depicted in Figure 7.1, displays existing control activities. Management efforts to encourage employees to uphold integrity, ethics and professional standards as the foremost important organization design depends on the following factors: protection of its inventory and intangible assets. As a manufacturing company, the inventory of raw material, work-in-process (WIP) and finished goods are the largest assets of the organization and the misappropriation of the inventory can negatively affect its financial standing. Moreover, in order for APCI's products to maintain a superior product compare to other competitors, it must safeguard its innovative ideas. Competitive pricing may determine the success of the organization. By preventing fraud, decreased cost can be achieved by APCI. Despite its current control activities, there are control discrepancies in its current internal control system due to human involvement. Therefore, the implementation of biometrics tools provides organizations with the automations useful in identification and/or verification of employees, thereby, eliminating human errors or opportunities for fraud (Rodgers et al., 2014). Therefore, the discrepancies in the internal control system may be improved by incorporating biometrics into its current internal control system (Figure 7.3).

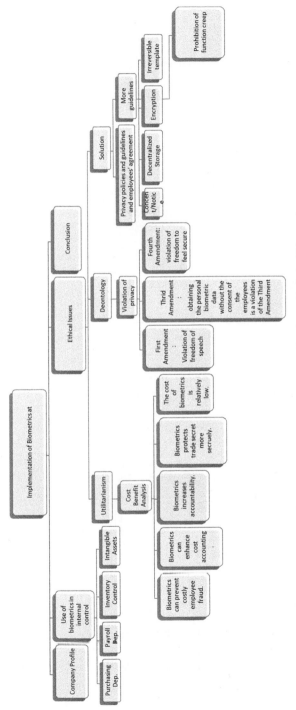

Figure 7.2. Flowchart of company's ethical standards and biometrics use

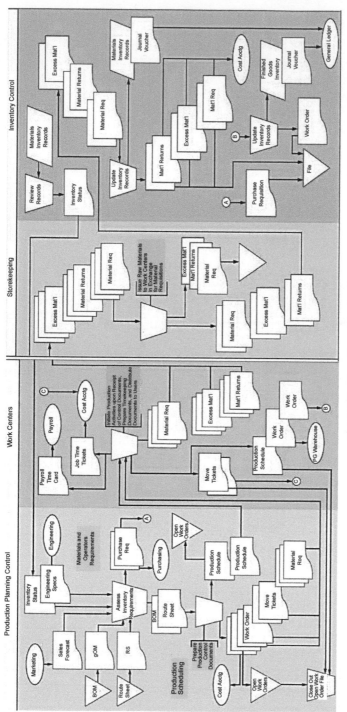

Figure 7.3. Operations flowchart

Bolle et al. (2004) defines biometrics as a science of identifying or verifying an identity of a person, based on physiological or behavioral characteristics. Biometrics can be utilized either for identification or verification purposes. Identification entails the process of recognizing a person from the list of people on the database (1:N matching), while verification involves the confirmation of the claimed identity (1:1 Matching) (Rodgers, 2011). Therefore, identification is more complex and requires more resources than verification. Verification of identity meets the need for organizational internal control in an effective manner; therefore, it is recommended for identification, in implementation of biometrics as an internal control measure.

1. Control activities analysis – Purchasing department

Currently, the purchasing department employs control features such as segregation of duties, accounting documentation, and processing regulation for its current control measures. After reviewing the current inventory status, engineering specifications, bill of materials (BOM), and the sales forecast drawn from suppliers' production schedule, production planning and process control department, APIC then completes the purchase requisition. Upon receipt of the purchase requisition, the purchasing department proceeds to order the requested raw materials. The segregation of duties between forecasting the quantity of raw material needed (authorization) and ordering (transaction processing) prevents excess ordering. In addition, purchase requisition provides the audit trail as an accounting document. Furthermore, for processing control purposes, accessing cards and passwords are utilized by purchasing agents to access the ordering program. This is performed since it is a preventative measure to deter unauthorized personnel to input unauthorized order.

Control discrepancy

Employees can assist one another when they are confronted with overload. Employees who are not purchasing agents help the purchasing agents when they are overloaded or are on vacation, by sharing the access cards and passwords in order to complete the purchasing process seamlessly. For this reason, purchasing agents leave their access cards and passwords at their workstation. Therefore, the intended controls can be circumvented when unauthorized personnel obtain the access card and passwords and process unauthorized purchase.

Recommendations

Biometrics can be introduced in the purchasing department for internal control purposes. Fingerprint scanning can replace access cards. The

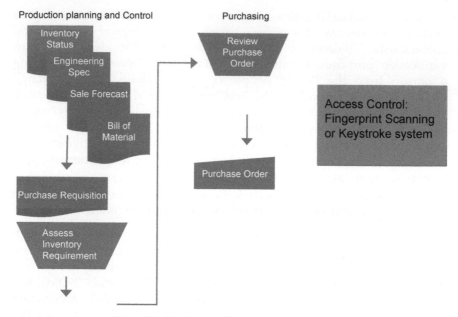

Figure 7.4. Purchasing department control activity

reason for its replacement is due to the fact that access cards can be shared, lost or stolen. Fingerprint scanning is easy to use and provides relatively high accuracy at a low price. In addition, the software for fingerprinting is typically less than $100 (Jain et al., 2004; Langenderfer and Linnhoff, 2005). Further, a keystroke system can be incorporated along with the use of fingerprint scanning. That is, everyone has a different typing dynamic (see Figure 7.4). Therefore, a perpetrator has a different typing rhythm and even when s/he uses the copied fingerprint to access the computer program. Also, his/her typing pattern will be different, and the system will not authorize the access. A keystroke system uses the existing keyboard and only have need of the software. Therefore, the cost to implement the keystroke system can be as low as less than $50 (Langenderfer and Linnhoff, 2005).

2. Inventory control—Work-in-process inventory

As a control measure, the transferred tickets are used to account for the work-in-process inventory. A supervisor at each work-center signs the transferred ticket to authorize the activities for or movement of each batch of products through the various work-centers.

Control discrepancy

Signature is a weak internal control measure as it can be easily forged by unauthorized personnel. The signature used for authorization purposes does verify the identity of the supervisor. Work-in-process (WIP) batch or products can be lost and stolen during the production process before completion.

Recommendations

Fingerprint scanning can be implemented in the process of authorizing the movement of the batch. As a unique biological characteristic, a fingerprint verifies the person who authorizes the movement is the person who he or she claims to be. It will prevent the circumvention of unauthorized movement of the WIP product inventory (Figure 7.5).

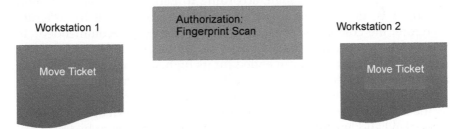

Figure 7.5. Work-in-process inventory control activity

3. Raw materials and finished good inventory

Current Control

Currently, two control measures are implemented in the system. The first control measure represents the physical access to raw material storeroom and finished goods warehouse. Further, security guards restrict the access only to the authorized personnel. Passing through the entry point, employees display their identification badges to gain the access. The second measure depicting segregation of duties between record-keeping (inventory control) and asset custody strengthens the internal control of the inventory.

Control discrepancy

A fake identification badge can be produced to gain unauthorized access to the warehouse or the storeroom. More serious issues can arise when there is collusion between the employees, the security guard and the inventory control management. Unauthorized personnel can gain access and misappropriate the inventory and inventory control management can purge the inventory record.

Recommendations

This well thought out crime through collusion can be circumvented by adopting facial recognition and incorporating it with the installation of scales to record the movement of inventory with identification of employees. Facial recognition has been widely utilized since it is relatively non-intrusive. Adding the facial recognition function to the existing security cameras can unobtrusively gather the biometrics information and control the entry to the area. The installed scales can measure the weight difference between the entry and the exit and detect and record the possession of inventory. Radio Frequency Identification (RFID) can be incorporated to enhance the detection of unauthorized possession of inventory upon exit. Although, according to Langenderfer and Linnhoff (2005), a false non-match rate (FNMR) of facial recognition is 10–20%, and therefore, the combined use of another biometrics is recommended. To address the low accuracy issue of face recognition, the integrated use of a fingerprint scanner is recommended for its high accuracy. Employees can gain access to the sensitive areas, such as the storeroom or warehouse by scanning their fingerprint and the possession of inventory can be detected by the difference in weights between the entry and the exit of an employee recognized and matched by the face recognition program or by the use of RFID (Figure 7.6).

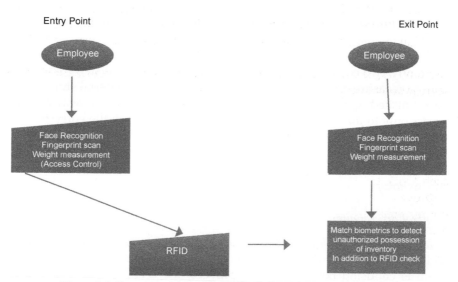

Figure 7.6. Raw materials and finished good inventory control activity

4. Payroll department

Current control

The current control dictates that employees sign in and out on the time sheets manually. As a control measure, supervisors review and reconcile the time sheets with actual attendance. Reviewed time card records are provided to payroll clerk for input. In addition, the personnel action form authorizes the addition and deletion of the employees. This procedure reduces payroll fraud when involving submitted time cards for bogus employees.

Control discrepancy

Supervisors' review and reconciliation of time and attendance records are not being followed, due to the amount of workload and many other responsibilities. A common practice of payroll fraud entails employees' rounding up hours or failing to report they are sick or on vacation, or personal days. "Buddy punching" is the most common fraud that employees commit without recognizing it as a fraud. Recent surveys reported that employees steal an average of 4.5 hours each week through tardy arrivals, early departures, and extended lunches or breaks (https://www.softwareadvice.com/hr/industryview/time-theft-report-2015/). Payroll fraud is not only about stealing from an organization, but also results in the incorrect measure of direct labor or indirect labor costs. As shown in the flowchart, the hours worked on the job are recorded on the job time ticket and affect the cost of each job. Therefore, false time record keeping can result in the incorrect cost of the products manufactured.

Management's overriding of controls or the collusion between the manager, time-keeping function and the personnel function can circumvent the existing control measure of transaction authorization and segregation of duties.

Recommendations

A fingerprint scanner can be implemented for timekeeping purposes as well as for cost accounting purposes by installing it at the entry point and each workstation. It is inexpensive and easy to install. Furthermore, it improves staff productivity and morale, reporting of accurate employee hours, scheduling, attendance, punctuality, and accountability. Face recognition using the security camera can be integrated with fingerprint scanning method to detect employees leaving work without signing out (Figure 7.7).

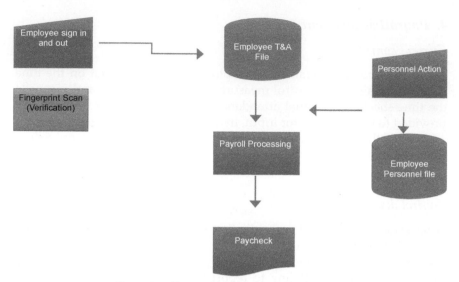

Figure 7.7. Time and attendance control activity

Current control

APCI has gained a competitive edge in its efficiency in production (trade secrets) through its intensive R&D program and the stored trade secrets on its database. For access control, the company has multilevel security to restrict access to trade secrets to a limited number of people and access cards and passwords that are used to access the computer network and database. In addition, the employees are required to sign the legally binding agreement that states their obligation to maintain the confidentiality of trade secrets during and at the time of leaving the employment.

Control discrepancy

Chandra and Calderon (2003) analyzed the different authentication methods and emphasized the vulnerability of the use of possession or knowledge for authentication purposes. Accordingly, possession-based authentication factors such as access cards, provide assurance that a user presents a valid card. Nonetheless, they do not offer direct assurance that an access cardholder, who is allowed access to an information system is indeed the person s/he claims to be. It only guarantees that a person who holds the access card has authorization to access. In addition, cards can be lost, stolen, shared, or duplicated (Chandra and Calderon, 2003).

Passwords that were used as a secondary authentication method, can be stolen, guessed, or shared. Passwords only provide assurance that the person at the keyboard knows the password. They do not offer assurance that the person at the keyboard is indeed the person he or she claims to be (Chandra and Calderon, 2003).

The agreement that employees sign at the beginning of their employment might be binding in court to collect damage from the breach of confidentiality of trade secrets. Nonetheless, damages are incurred when an organization loses its competitive advantages in the marketplace. That is, once the trade secrets are stolen and known to the public including the competitors, it is not recoverable.

Recommendations

A keyboard with a fingerprint scanner can be adopted for access control to replace the access cards and passwords. To reinforce the authentication process, facial recognition or iris scan can be combined with the fingerprint scanning. Both the facial recognition and the iris scan require the use of camera technology. As the computer monitors are equipped with a high definition camera, the overall organizational cost can be reduced in the implementation of either method. Other costs to the system would include software programming and maintenance fora facial recognition program or an iris scan program. Iris scan is higher in cost than face recognition; however, iris scan is more effective and accurate in the verification process than face recognition. Safeguarding its trade secrets is very important for organizations to be well positioned in the market due to its innovative technology. Once the trade secrets are stolen and become public, companies lose their competitiveness.

The implementation of biometrics will not only enhance internal control activities, but also improve communication with management on the weakness or potential risks residing in the organization by providing regular or ad-hoc reports. In addition, monitoring the internal control can be facilitated with the implementation of biometrics as it provides substantial evidences in the reports.

Ethical Issues

Biometrics is a double-edged sword that brings about the dilemma between the benefits and the violation of individual rights. Advantages and disadvantages of the implementation of biometrics at workplaces can be weighed using both utilitarianism and deontology.

Utilitarianism

Utilitarianism is concerned with consequences and focuses on the greatest good for the greatest number of people. Based on the available information

(I) the judgment is made to make an ethical decision (D) (Rodgers and Gago, 2001). Utilitarianism compares cost and benefits (cost-benefit analysis). In a utilitarian point of view, biometrics is ethical because it offers many benefits. First of all, biometrics can prevent costly employee fraud. According to Guffey and West (1996), 30% of all business failure is due to employee theft and the total cost of business fraud amounts to $50 billion annually. These costs are absorbed by the public in the form of higher prices and higher taxes. When biometrics is implemented in the workplace, these costs can be eliminated by preventing the misappropriation and reducing the time and attendance fraud, which saves consumers billions of dollars.

In addition to the cost savings, biometrics aided by "big data" can enhance cost accounting. The organization that implements biometrics can gain a competitive edge by pricing their products accurately. The use of biometrics for time and attendance does not only reduce the downtime, but also provides the accurate time spent for manufacturing. This provides the organization with the real-time actual cost of the product.

Furthermore, monitoring and recording increases accountability. Information gathered from the biometric information can be used in other algorithmic pathways (see Figure 7.1) for improved efficiencies of the organization. Increased accountability enforces compliance with policies and procedures. This can result in enhanced accountability are improved quality and punctuality.

In addition, trade secrets can be protected more securely in the organization. Trade secrets are paramount to the survival of a business and, biometrics reinforces the protection of these trade secrets. Trade secrets are not recoverable once they are publicized. Therefore, the prevention of the trade secret embezzlement is vital to the survival of a business and biometrics provides one of the strongest security measure in terms of the access control and monitoring.

Compared to the benefits that biometrics provides, its cost is relatively low. As discussed previously, a fingerprint scanner and the software has an estimated cost about $100 and the keystroke analysis software is about $50. The cost of biometric devices has decreased and predicted to further decrease. Moreover, employee ID cards, the alternative to biometrics, are not permanent and therefore, the costs are incurred periodically. In conclusion, biometrics along with "big data" can be used as an utilitarian ethical Artificial Intelligence choice as it benefits the greatest number of people at the minimum costs.

On the other hand, deontology focuses on adherence to independent moral rules or duties (P) and ignores additional information (I) in judging (J) to make an ethical decision (D) (Rodgers and Gago, 2001). Constitutional rights, such as freedom of speech and the right to be secure are protected under certain rights that are part of quite a few nations. Therefore, the companies operating in the US have duties to protect these rights to abide by the constitution.

Deontology

From a deontological viewpoint, biometric tools may violate the privacy of employees by monitoring and recording the movement and the conversation of employees continuously as well as by collecting and storing employee's unique biometric data. It can be construed as a violation of freedom of speech, which was granted by the First Amendment of Bill of Rights (Rodgers, 2010, 2012). Biometrics implemented in a certain manner may also violate the Third Amendment by obtaining the personal biometric data without the consent of the employees. The biggest argument of the violation of constitutional rights lies on the Fourth Amendment. The Fourth Amendment states, "the right of the people to be secure in their persons, houses, papers, and effects, against unreasonable searches and seizures, shall not violated..." and employees do not feel secure by being monitored constantly and having their personal biometric data collected and stored by the employer. Hence, from a narrow deontological viewpoint, biometrics may be viewed as unethical. This type of interpretation is due to the fact that certain employment of biometrics may infringe upon the constitutional rights under the First, Third and Fourth Amendments to the Bill of Rights (Rodgers, 2010, 2012).

Solutions

The utilitarian and deontological competing ethical viewpoint on biometrics can be reconciled by employing a few precautionary measures. First of all, employers can notify employees of the use of biometrics for electronic monitoring and access control prior to the implementation of biometric tools through the company privacy policies and guidelines and have employees sign the agreement. A landmark Supreme Court case, Katz vs. United States, 389 US 347(1967) established the key legal test for what counts as reasonable search under the Fourth Amendment: "what a person knowingly exposes to the public is not a subject of Fourth Amendment protection." Therefore, when employees are aware of and provided the consent to the collection and the use of biometrics, biometrics is no longer an unreasonable search. According to Nord et al. (2006), the Fourth Amendment is limited to protecting federal, state, and municipal employees in most cases.

Private sector employees can seek the protection of their rights under the privacy issues (Rodgers, 2010, 2012) or Electronic

Communications Privacy Act (ECPA) (https://it.ojp.gov/PrivacyLiberty/authorities/statutes/1285). Nonetheless, ECPA has three exceptions: provider exception, ordinary course of business, and consent exception. The implementation of biometrics accompanied by employees' agreement can avoid the legal liability to the employees under ECPA as it satisfies all the exception rules. In addition, employees' agreement can also prevent lawsuits based on the Second Amendment as employees have given the employer their consent to obtain individuals' biometrics. The freedom of speech under the First Amendment is also legally covered if the organization establishes the privacy guidelines, stating that the only work-related conversations are monitored.

Mere avoidance of the legal liability does not indicate that the use of biometrics is an ethical choice. The ethical issues regarding the implementation of biometrics may potentially lead to data misuses such as identity theft, fraud and function creep (Cavoukian et al., 2012). Langenderfer and Linnhoff proposed the following guidelines to reduce the data misuses and the intrusion of privacy (2005).

1. No biometric data are collected without notice or consent.
2. Biometric data are stored in a decentralized manner such as on a smart card rather than in central storage.
3. Biometric data are encrypted.
4. Biometric data are stored in irreversible template format rather than a photograph or a complete scan.
5. Sharing biometric data with other entities is prohibited.

Conclusion

Buttressing the "intelligent" aspect of Artificial Intelligence, advances in technology have led to the development of machine learning to make predictions or decision choices without being explicitly programmed to accomplish a task. With machine learning, algorithms and statistical models permit systems to "learn" from data, and make decisions, relying on patterns and inference instead of specific instructions.

Unfortunately, the possibility of fashioning machines that can think results in a myriad of ethical issues. From pre-existing biases utilized to train Artificial Intelligence to social manipulation by means of newsfeed algorithms and privacy invasions via facial or iris recognition, ethical issues are harvesting as Artificial Intelligence continues to expand in prominence and utilization. This notion focuses on the need for reasonable conversation surrounding how we can responsibly assemble and embrace these technologies.

The biggest ethical challenge confronting global society can be shaped by better understanding six dominant ethical algorithmic pathways as applied to Artificial Intelligence. To work within the global society, Artificial Intelligence has to be aware of the nuances and particulars of specific societies as they relate to the six ethical positions. An Artificial Intelligence system in a high surveillance country might differ from its equivalents in other parts of the world. Then, of course, there is ethical divide within societies.

The Artificial Intelligence policy landscape is still in its infancy. Many policymakers see this moment as the beginning of an Artificial Intelligence arms race in need of public funding and deregulation. Others are calling for comprehensive guidelines that address ethical algorithms, workforce retraining, public safety, antitrust, and transparency.

Biometrics can improve the internal control by reducing the human involvement; however, it also imposes the concern on the violation of privacy. Many people have opposed the implementation of biometrics for the violation of constitutional rights and privacy issues. Yet, the answer does not have to be black and white. Society in the 21st century can enjoy the benefit of technology such as biometrics while respecting the rights and privacy by establishing policies, procedures and guidelines for the appropriate use of this advanced technology can enhance its internal control by implementing the suggested biometric tools along with the policies such as notification and consent, biometric encryption, decentralized data storage, use of templates, and prohibition of function creep.

To sum up, as an increasing number of Artificial Intelligence enabled devices are developed and utilized by society and organizations around the world, the need to keep those devices secure has never been more essential. Artificial Intelligence's increasing capabilities and utilization dramatically enhance the opportunity for nefarious uses. Consider the threatening possibility of autonomous vehicles and weapons such as armed drones falling under the control of disgruntled employees.

Due to the aforementioned caveats, it has become critical that information technology departments, consumers, suppliers, business leaders and the government, fully understand cybercriminal strategies that could lead to an Artificial Intelligence-driven threat to society. If they don't, maintaining the security of these traditionally insecure devices and protecting an organization's digital transformation becomes a nearly unachievable endeavor.

References

Bolle, R.M., Connell, J.H., Pankanti, S., Ratha, N.K. and Senior, A.W. 2004. Guide to Biometrics. New York: Springer-Verlag.

Cavoukian, A., Chibba, M. and Stonianov, A. 2012. Advance in biometric encryption: taking privacy by design from academic research to deployment. Review of Policy Research, 29(1): 37–61.

Chandra, A. and Calderon, T.G. 2003. Toward a biometric security layer in accounting systems. Journal of Information Systems, 17(2): 51–70.

Guffey, C.J. and West, J.F. 1996. Employee privacy: legal implication for managers. Labor Law Journal, 743.

Katz v. United States, 389 U.S. 347(1967).

Langenderfer, J. and Linnhoff, S. 2005. The emergence of biometrics and its effect on consumers. The Journal of Consumer Affairs, 39(2): 314–338.

Rodgers, W. and Gago, S. 2001. Cultural and ethical effects on managerial decisions: examined in a throughput model. Journal of Business Ethics, 31: 355–367.

Rodgers, W. 2003. Measurement and reporting of knowledge-based assets. Journal of Intellectual Capital, 4: 181–190.

Rodgers, W. 2007. Problems and resolutions to future knowledge-based assets reporting. Journal of Intellectual Capital, 8: 205–215.

Rodgers, W. 2009. Ethical Beginnings: Preferences, Rules, and Principles Influencing Decision Making. NY: iUniverse, Inc.

Rodgers, W. 2010. E-commerce and Biometric Issues Addressed in a Throughput Model. Hauppauge, NY: Nova Publication.

Rodgers, W. 2012. Biometric and Auditing Issues Addressed in a Throughput Model. Information Age Publishing.

Rodgers, W., Söderbom, A. and Guiral, A. 2014. Corporate social responsibility enhanced control systems reducing the likelihood of fraud. Journal of Business Ethics, 91(Supplement 1): 151–166.

The National AI R&D Strategic Plan. 2016. National Science and Technology Council, USA. https://www.nitrd.gov/pubs/national_ai_rd_strategic_plan.pdf.

Tillquist, J. and Rodgers, W. 2005. Valuation of information technologies in firms: Asset scope and asset specificity. Communications of the ACM, 48: 75–80.

8

Conclusions

"The success of cognitive computing will not be measured by Turing tests or a computer's ability to mimic humans. It will be measured in more practical ways, like return on investment, new market opportunities, diseases cured, and lives saved."

—Dr. John Kelly III, IBM Senior Vice President for Research and Solutions

"The bottom 90 percent, especially the bottom 50 percent of the world in terms of income or education, will be badly hurt with job displacement... The simple question to ask is, 'How routine is a job?' And that is how likely [it is] a job will be replaced by Artificial Intelligence, because Artificial Intelligence can, within the routine task, learn to optimize itself. And the more quantitative, the more objective the job is—separating things into bins, washing dishes, picking fruits and answering customer service calls—those are very much scripted tasks that are repetitive and routine in nature. In the matter of five, 10 or 15 years, they will be displaced by Artificial Intelligence."

—AI oracle and venture capitalist Dr. Kai-Fu Lee

Artificial Intelligence has become increasingly more pervasive in our everyday life. Many benefits are leaping forward in terms of scientific progress, human well-being, economic value, and the possibility of exploring solutions to major social and environmental problems. Nonetheless, such a commanding technology also raises some concerns, such as its capability to make imperative decisions in a manner that individuals would perceive as ethical issues.

Artificial Intelligence algorithms provide the framework to explain its reasoning and decision-making. To be knowledgeable about Artificial Intelligence is to be aware of ethical considerations, which is aligned to human values that are relevant to the problems being tackled, Since many successful Artificial Intelligence techniques rely on huge amounts of data, it is important to know how data are handled by Artificial Intelligence systems and by those who produce them.

Moreover, machine learning, neural networks and deep learning have arrived in this millennium to power Artificial Intelligence systems. The Artificial Intelligence information revolution is increasing at a huge pace, and, therefore society must be wary of the implications that come along with it. Further, Artificial Intelligence can provide us with how long our commute will be, what music we should listen to, and what content we would likely engross ourselves with. Nevertheless, while users are scrolling through their webpage newsfeed, an algorithm somewhere could be determining their medical diagnoses, their parole eligibility, or their career prospects.

Machine learning implements algorithms to parse data, learn from that data, and make informed decision choices based upon what it has learned. Deep learning structures algorithms work in layers to generate an artificial "neural network" that can learn and make intelligent decision choices on its own. Deep learning can be viewed as a subfield of machine learning. While both fall under the broad category of Artificial Intelligence, deep learning is typically what is the motivator for most human-like Artificial Intelligence.

The rapid pace of technological innovations has soared. Further, "disruption" has been the watchword of the last decade. Technological advancements have been jet streaming in the beginning of the third millennium. This is the time period of the Common Era in the Gregorian calendar in the current millennium spanning the years 2001 to 3000 (i.e., 21st to 30th centuries). In this period, Artificial Intelligence is leading the way by simulating human intelligence processes by machines, especially computer systems. These processes embrace learning (the acquisition of information and rules for using the information), reasoning (using rules to obtain approximate or definite conclusions) and self-correction.

Artificial Intelligence is transforming the way we view our environment. Artificial Intelligence "robots" are everywhere. From our phones to devices like Amazon's Alexa, we live in a world surrounded by machine learning, neural networks and deep learning.

In addition, deep learning systems are those that depend upon non-linear neural networks to build machine learning systems, often relying upon using the machine learning to actually model the system doing the modeling. It can be described in some quarters as a subset of machine learning with a specific emphasis on neural nets.

Moreover, sophisticated Artificial Intelligence system is one that can learn on its own. For example, neural networks similar to Google's

DeepMind, are making connections and reach meanings without relying on pre-defined behavioral algorithms. Artificial Intelligence is constantly improving upon past iterations, getting smarter and more aware, allowing it to enhance its capabilities and its knowledge.

Artificial Intelligence technology's objective is to replicate or surpass abilities (in computational systems) that would require 'intelligence' if individuals were to perform them. These include learning and adaptation; sensory understanding and interaction; reasoning and planning; search and optimization; autonomy; and creativity. The applications of Artificial Intelligence systems, embracing but not limited to machine learning, are various, ranging from understanding healthcare data to autonomous and adaptive robotic systems, to smart supply chains, video game design and content creation. This research area primarily envelops fundamental enhancements in Artificial Intelligence technologies, while applications of such technologies are portrayed within other subject domains.

Examples of Artificial Intelligence progress and significant strides include but is not limited to the following areas. Suggestions include what restaurants that people should dine. That is, Artificial Intelligence helpers such as Siri, Google Assistant, Microsoft Cortana, Amazon Alexa assist in this endeavor. Further, tasks such as the autocorrect function in the smartphone keyboard and the automated tagging functions on Facebook are all coordinated by the power of Artificial Intelligence. In other words, Artificial Intelligence tools are attempting to make computers mimic human intelligence, which is aided by machine learning, neural networking, and deep learning technology. For example, in neural networking, the process is to make transistors behave like neurons of the human brain. Machine learning applies to the usage of artificial neural networks (ANNs) in order to facilitate learning at multiple layers. Deep learning is based on data presentation rather than task-based algorithms. Although the future of Artificial Intelligence may permit machines to make decision choices like people, the present is already influencing human decisions, especially for organizations.

Artificial Intelligence Impact on Our Lives

Since Alan Turing's paper in the 1950s, there has been a debate on what Artificial Intelligence can do and how people will be influenced by its methods. Overall, this thought process and speculation are not surprising. Artificial Intelligence tools are characteristic in the case of any evolving field whereby complete knowledge is yet to be acquired. The only difference is that Artificial Intelligence will continuously evolve and, therefore, being able to foresee the subsequent change becomes a big question.

In such an environment, developments could be set in motion contingent on needs and not always the other way around. In other words, generating a need and then using an innovation following its development. A fair amount of collaboration among academia, the public and private sectors becomes essential. This encounter will embolden innovation in an effective and efficient manner. Constant interchange among these three pillars (i.e., academia, public sector and private sector) will impede the rare possibility of an innovation being at odds with human interest.

The Artificial Intelligence continuum comes to the forefront with Artificial Intelligence innovations that fall under amplified, aided and autonomous intelligence assisting users understand and make decision choices regarding which level of intelligence is beneficial and is a pre-requisite for taking action. This will make the acceptance of Artificial Intelligence simpler to the multitudes. At the same time, the continuum could be implemented to understand economic ramifications, complexity of use and decision-making implications. While academia and the private sector conduct research on various Artificial Intelligence problems with diverse implications in mind, the public sector, with its assortment of initiatives can identify areas where segments of the Artificial Intelligence continuum can be used to escalate reach, effectiveness and efficiency, thereby assisting in providing direction to the innovative Artificial Intelligence research prevalent in the global community.

A collaborative innovation environment with regular dialogue among academia, private and public sectors will assist in recognizing novel areas and operations among the population. For example, Artificial Intelligence may be implemented to offer universal and proactive advisory conveyance to the population through public call centers, linking information from a range of government sources. Meanwhile, the plethora of big data produced from the interactions in the aforementioned process and other digital initiatives can be utilized to develop leading-edge conclusions. Collaboration among the three pillars may also assist in dealing with a comprehensive view of problems. This could very well lead to intelligent and innovative ways to enhance the efficiency and effectiveness of services delivered to society. Likewise, these can be viewed as it is applied through an Artificial Intelligence continuum which will help provide the benefits of Artificial Intelligence evolution to the global community according to their needs.

Artificial Intelligence is likely to become a progressively prevailing aspect of our world and understanding the future of Artificial Intelligence and its impact on future society are critically imperative for practice and research areas. Understanding of Artificial Intelligence decisions is vital, as is the use of Artificial Intelligence to simplify data in order to expedite human understanding and decision-making.

Where do we Go from Here?

A national policy could be helpful in clearly defining standards and benchmarks that can be effectively implemented to gauge progress in Artificial Intelligence innovation and commercialization in a host of application domains. By nature, the Artificial Intelligence space is an indeterminate and vague—one with no direct traceability of returns from investment in innovation and capability building. This makes it all the more essential for transitional tangible progress to be measured against set targets from time to time.

Although, Artificial Intelligence technologies provide a range of new functionality for organizations, the use of Artificial Intelligence raises certain ethical questions. For example, deep learning algorithms that underline many of the most advanced Artificial Intelligence tools, are only as smart as the data they are given in training. For the reason that people select what data should be utilized for training an Artificial Intelligence program, the potential for human bias is inherent and must be monitored closely.

Moreover, a resilient presence in Artificial Intelligence research and development is a precondition for a nation to gain a lead in an automation-driven future. For this, the national policy ought to take precise stock of current and future demand for Artificial Intelligence experts and affiliated areas. Building expertise, on the other hand, will necessitate governments to gauge the current educational corridors and curricula and, if required, renovate and refurbish the same to make available skill upgradation initiatives for a workforce that strives to stay relevant in a fast-evolving technology landscape.

To institute control over Artificial Intelligence's data, algorithms, processes, and reporting frameworks, a system requires combined teams of technical, business and internal audit specialists. These teams will have to consider appropriate trade-offs since they continually test and monitor controls. The issue of interpretability looms large, which should be a blend of performance, cost, need, and the extent of individuals' expertise involved in the system. For example, self-driving automobile, an Artificial Intelligence healthcare diagnosis, and an Artificial Intelligence-led marketing campaign requires different levels and types of interpretability and related controls.

The Throughput Model's six dominant algorithms can help explain how Artificial Intelligence learns with better data governance. Moreover, some organizational issues will have Artificial Intelligence solutions that will necessitate training data that organizations may not have available. Nonetheless, novel and augmented machine learning and deep learning techniques can enable Artificial Intelligence to construct its own data based on a few samples. They can also transfer models from one task with a great deal of data to another one that lacks data. Artificial Intelligence can oftentimes synthesize its own training data by using techniques such as reinforcement learning, active learning, generative and adversarial networks. Simulations established on probabilities can also produce "synthetic" data that can be implemented to train Artificial Intelligence.

Artificial Intelligence Types of Knowledge Engineering

Autonomous intelligence

Adaptive/continuous systems that take over decision making in some cases. But they will do so only after the human decision maker starts trusting the machine or becomes a liability for fast transactions. In this type of intelligence, the decision rights are with the machine and hence it is fundamentally different from assisted intelligence.

Augmented intelligence

Enhancing human ability to do the same tasks faster or better. Humans still make some of the key decisions, but Artificial Intelligence executes the tasks on their behalf. The decision rights are solely with humans.

Assisted intelligence

Humans and machines learn from each other and redefine the breadth and depth of what they do together. Under these circumstances, the human and the machine share the decision rights.

Currently, machines are able to process data far more quickly than we can. Nonetheless, as human beings, we have the capability to think abstractly, strategize, and tap into our thoughts and memories to make informed decisions or come up with creative ideas. This type of intelligence makes us superior to machines, but it is difficult to define since it is primarily driven by our capability to be sentient beings. Consequently, it is something that is very hard to replicate in machines. In future, "general" Artificial Intelligence is expected to be able to reason, solve problems, make judgements under uncertainty, plan, learn, integrate prior knowledge in decision-making, and be innovative, imaginative and creative. In other words, any intellect that vastly exceeds the cognitive performance of people in virtually all domains of interest will be deemed as super intelligence.

In future, this type of intelligence may surpass human intelligence in all aspects from creativity, to general wisdom, to problem-solving. Machines may be capable of displaying intelligence that we have not seen in the most brilliant amongst us. This is the type of Artificial Intelligence that many people are concerned about, and the type of Artificial Intelligence that people think will lead to an alteration of the human species. Nonetheless, invoking ethical algorithms into Artificial Intelligence systems may arrest the fears that some people have regarding advanced Artificial Intelligence machinery.

Looking at Artificial Intelligence this way raises the question of whether there are any skills that remain distinctive for humans and out of reach of Artificial Intelligence. This question is difficult to answer given the incredible progress Artificial Intelligence has experienced in the past. In spite of everything, it appears likely that people will always have an upper hand where artistic creativity is concerned. Essentially, Artificial Intelligence is based on pattern recognition or curve fitting (i.e., finding a relationship that explains an existing set of data points), while "creativity is intelligence having fun," as Albert Einstein put it. At this stage, it seems it will be a while until Artificial Intelligence systems will be able to solve truly creative tasks.

The vital social issue enveloping biometrics is the seemingly irrevocable link between biometric traits and a consistent information record about an individual. Unlike most other forms of recognition, biometric techniques are stalwartly fixed to our physical bodies. The close-fitting link between personal records and biometrics can have both positive and negative consequences for individuals and for society at large. Convenience, improved security, and fraud diminution are some of the benefits sometimes connected with the utilization of biometrics. Those benefits may flow to certain individuals, organizations, and societies; however, they are sometimes realized only at the expense of others. Who benefits at whose expense and the relative balance between benefits and costs can impact on the success of biometric deployments.

In summary, Artificial Intelligence technology is not the end but only a means towards effectiveness and efficiency, improved innovative capabilities, and better opportunities. Further, we have witnessed the employment of Artificial Intelligence in several industries that have begun to adopt these systems into their operations. Moreover, Artificial Intelligence will also become smarter, faster, more fluid and human-like thanks to the inevitable rise of quantum computing. Quantum computers will solve most of life's most difficult problems and doubts affecting the environment, aging, disease, war, poverty, famine, the origins of the universe and deep-space exploration, etc. To conclude, it may eventually enhance all of our Artificial Intelligence systems, performing as the brains of these super-human machines.

Index

About the Author

Waymond Rodgers is a C.P.A. (inactive) and holds a Chair Professorship in the School of Business at the University of Hull (UK) and is a professor of Accounting and Information Systems at the University of Texas, El Paso (USA). Previously he was a professor at University of California, Riverside and Irvine. His degrees are from Michigan State University (B.A.), University of Detroit-Mercy (M.B.A.), University of Southern California, Ph.D. in accounting information systems; and an experimental psychology post-doctorate from the University of Michigan. He has received numerous research grants including from the National Science Foundation, Ford Foundation and Citibank. He is a Ford Foundation Fellow and received a Franklin Fellowship from the US State Department for his work on coherent sustainable and knowledge transfer strategy systems.

Dr. Rodgers recently received the Ethics Research Symposium Best Research Paper Award for his paper "*Artificial Intelligence Algorithmic Approach in Enhancing Auditors' Fraud Risk.*" He was working as an auditor with Ernst & Young and PriceWaterhouse Coopers, and was a commercial loan officer with Union Bank. Professor Rodgers' has published ten books including on decision-making, knowledge management, ethical and trust-based cyber security systems, as well as the use of biometric devices as a way of intensifying identification and authentication of cyber control systems. Professor Rodgers has been published in leading international journals such as *Accounting Forum, Auditing: A Journal of Practice & Theory, European Accounting Review, Journal of Business Ethics, Journal of the Association of Information Systems, Journal of World Business, Management Learning, Management Science, Organization Studies, Sustainability* among others.